T0192220

Communications
in Computer and Information Science   **1784**

Rationale

The CCIS series is devoted to the publication of proceedings of computer science conferences. Its aim is to efficiently disseminate original research results in informatics in printed and electronic form. While the focus is on publication of peer-reviewed full papers presenting mature work, inclusion of reviewed short papers reporting on work in progress is welcome, too. Besides globally relevant meetings with internationally representative program committees guaranteeing a strict peer-reviewing and paper selection process, conferences run by societies or of high regional or national relevance are also considered for publication.

Topics

The topical scope of CCIS spans the entire spectrum of informatics ranging from foundational topics in the theory of computing to information and communications science and technology and a broad variety of interdisciplinary application fields.

Information for Volume Editors and Authors

Publication in CCIS is free of charge. No royalties are paid, however, we offer registered conference participants temporary free access to the online version of the conference proceedings on SpringerLink (http://link.springer.com) by means of an http referrer from the conference website and/or a number of complimentary printed copies, as specified in the official acceptance email of the event.

CCIS proceedings can be published in time for distribution at conferences or as post-proceedings, and delivered in the form of printed books and/or electronically as USBs and/or e-content licenses for accessing proceedings at SpringerLink. Furthermore, CCIS proceedings are included in the CCIS electronic book series hosted in the SpringerLink digital library at http://link.springer.com/bookseries/7899. Conferences publishing in CCIS are allowed to use Online Conference Service (OCS) for managing the whole proceedings lifecycle (from submission and reviewing to preparing for publication) free of charge.

Publication process

The language of publication is exclusively English. Authors publishing in CCIS have to sign the Springer CCIS copyright transfer form, however, they are free to use their material published in CCIS for substantially changed, more elaborate subsequent publications elsewhere. For the preparation of the camera-ready papers/files, authors have to strictly adhere to the Springer CCIS Authors' Instructions and are strongly encouraged to use the CCIS LaTeX style files or templates.

Abstracting/Indexing

CCIS is abstracted/indexed in DBLP, Google Scholar, EI-Compendex, Mathematical Reviews, SCImago, Scopus. CCIS volumes are also submitted for the inclusion in ISI Proceedings.

How to start

To start the evaluation of your proposal for inclusion in the CCIS series, please send an e-mail to ccis@springer.com.

Shiyu Yang · Saiful Islam
Editors

Web and Big Data

APWeb-WAIM 2022 International Workshops

KGMA 2022, SemiBDMA 2022, DeepLUDA 2022
Nanjing, China, November 25–27, 2022
Proceedings

Springer

Editors
Shiyu Yang
Guangzhou University
Guangzhou, China

Saiful Islam 🔟
Griffith University
Gold Coast, QLD, Australia

ISSN 1865-0929   ISSN 1865-0937 (electronic)
Communications in Computer and Information Science
ISBN 978-981-99-1353-4   ISBN 978-981-99-1354-1 (eBook)
https://doi.org/10.1007/978-981-99-1354-1

This Springer imprint is published by the registered company Springer Nature Singapore Pte Ltd.
The registered company address is: 152 Beach Road, #21-01/04 Gateway East, Singapore 189721, Singapore

Preface

The Asia Pacific Web (APWeb) and Web-Age Information Management (WAIM) Joint International Conference on Web and Big Data (APWeb-WAIM) is a leading international conference for researchers, practitioners, developers, and users to share and exchange their cutting-edge ideas, results, experiences, techniques, and applications in connection with all aspects of web and big data management. The conference invites original research papers on the theory, design, and implementation of data management systems. As the 6th edition in the increasingly popular series, APWeb-WAIM 2022 was held in Nanjing, China, November 25–27, 2022. Along with the main conference, the APWeb-WAIM workshops provide an international forum for researchers to discuss and share their pioneered works.

The APWeb-WAIM 2022 workshop volume contains the papers accepted for three workshops held in conjunction with APWeb-WAIM 2022. The three workshops were selected after a public call-for-proposal process, and each of them had a focus on a specific area that contributed to the main themes of the APWeb-WAIM conference. After the single-blinded review process, out of 39 submissions, the three workshops accepted a total of 23 papers, marking an acceptance rate of 58.97%. Each submission has been peer-reviewed by 3–4 reviewers. The three workshops were as follows:

- The Fifth International Workshop on Knowledge Graph Management and Applications (KGMA 2022)
- The Fourth International Workshop on Semi-structured Big Data Management and Applications (SemiBDMA 2022)
- The Third International Workshop on Deep Learning in Large-scale Unstructured Data Analytics (DeepLUDA 2022)

As a joint effort, all organizers of APWeb-WAIM conferences and workshops, including this and previous editions, have made APWeb-WAIM a valuable trademark through their work. We would like to express our thanks to all the workshop organizers and Program Committee members for their great efforts in making the APWeb-WAIM 2022 workshops such a great success. Last but not least, we are grateful to the main conference organizers for their leadership and generous support, without which this APWeb-WAIM 2022 workshop volume wouldn't have been possible.

December 2022

Shiyu Yang
Saiful Islam

Organization

APWeb-WAIM 2022 Workshop Co-chairs

Shiyu Yang   Guangzhou University, China
Saiful Islam   Griffith University, Australia

APWeb-WAIM 2022 Publication Chairs

Chuanqi Tao   Nanjing University of Aeronautics and
                               Astronautics, China
Lin Yue   University of Newcastle, Australia
Xuming Han   Jinan University, China

KGMA 2022

Workshop Chairs

Xin Wang   Tianjin University, China
Qingpeng Zhang   City University of Hong Kong, China
Yunpeng Chai   Renmin University of China, China

Program Committee

Huajun Chen   Zhejiang University, China
Wei Hu   Nanjing University, China
Saiful Islam   Griffith University, Australia
Jiaheng Lu   University of Helsinki, Finland
Jianxin Li   Deakin University, Australia
Ronghua Li   Beijing Institute of Technology, China
Jeff Z. Pan   University of Aberdeen, UK
Jijun Tang   University of South Carolina, USA
Haofen Wang   Tongji University, China
Hongzhi Wang   Harbin Institute of Technology, China
Junhu Wang   Griffith University, Australia
Meng Wang   Southeast University, China

Xiaoling Wang	East China Normal University, China
Xuguang Ren	G42 Inception Institute of Artificial Intelligence, UAE
Guohui Xiao	Free University of Bozen-Bolzano, Italy
Zhuoming Xu	Hohai University, China
Qingpeng Zhang	City University of Hong Kong, China
Xiaowang Zhang	Tianjin University, China

SemiBDMA 2022

Workshop Chairs

Baoyan Song	Liaoning University, China
Xiaoguang Li	Liaoning University, China
Linlin Ding	Liaoning University, China
Yuefeng Du	Liaoning University, China

Program Committee

Ye Yuan	Northeastern University, China
Xiangmin Zhou	RMIT University, Australia
Bo Ning	Dalian Maritime University, China
Yongjiao Sun	Northeastern University, China
Yulei Fan	Zhejiang University of Technology, China
Guohui Ding	Shenyang Aerospace University, China
Bo Lu	Dalian Nationalities University, China
Linlin Ding	Liaoning University, China
Yuefeng Du	Liaoning University, China

DeepLUDA 2022

Organizers

Tae-Sun Chung	Ajou University, South Korea
Qiang Liu	Beijing Institute of Petrochemical Technology, China
Rize Jin	Tiangong University, China

Workshop Chair

Joon-Young Paik — Tiangong University, China

Program Committee

Liangfu Lu	Tianjin University, China
Yunbo Rao	University of Electronic Science and Technology of China, China
Weiwei Yang	General Data Technology Co., Ltd., China
Se Jin Kwon	Kangwon National University, South Korea
Caie Xu	Zhejiang University of Science & Technology, China
Ziyang Liu	Kyonggi University, South Korea
He Li	Xidian University, China
Yenewondim Biadgie	Ajou University, South Korea
Muhammad Attique	Sejong University, South Korea
Zhen Wang	Tianjin University of Finance and Economics, China
Huayan Zhang	Tiangong University, China
Gaoyang Shan	Ajou University, South Korea
Xun Luo	Tianjin University of Technology, China
Guanghao Jin	Beijing Polytechnic University, China
Jing Yang	Guangdong University of Foreign Studies, China
Yanji Piao	Yanbian University, China
Botao Guo	Chongqing Zhongke Automotive Software Innovation Center, China
Zhen Zhang	Tiangong University, China
Qianjin Guo	Beijing Institute of Petrochemical Technology, China

Contents

**The Third International Workshop on Deep Learning in Large-Scale
Unstructured Data Analytics**

The Fifth International Workshop on Knowledge Graph Management and Applications

Knowledge-Enhanced Medical Visual Question Answering: A Survey (Invited Talk Summary)

Haofen Wang[(✉)] and Huifang Du

Tongji University, Shanghai, China
`carter.whfcarter@gmail.com, duhuifang@tongji.edu.com`

Abstract. Medical Visual Question Answering (Med-VQA) is a task in the field of Artificial Intelligence where a medical image is given with a related question, and the task is to provide an accurate answer to the question. It involves the integration of computer vision, natural language processing, and medical domain knowledge. Furthermore, incorporating medical knowledge in Med-VQA can improve the reasoning ability and accuracy of the answers. While knowledge-enhanced Visual Question Answering (VQA) in the general domain has been widely researched, medical VQA requires further examination due to its unique features. In the paper, we gather information on and analyze the current publicly accessible Med-VQA datasets with external knowledge. We also critically review the key technologies combined with knowledge in Med-VQA tasks in terms of the advancements and limitations. Finally, we discuss the existing challenges and future directions for Med-VQA.

Keywords: Medical visual question answering · Medical knowledge · Computer vision · Natural language processing

1 Introduction

Medical Visual Question Answering (Med-VQA) is an emerging field that combines computer vision, natural language processing, and healthcare, aiming to enable AI systems to understand and answer questions based on medical images. These images can be in the form of X-rays, CT scans, MRIs, pathology, etc. With the increasing availability of medical imaging data, advanced technologies like pre-trained models [1,2] and few shot learning [3,4], as well as the growing need for efficient and accurate disease diagnosis [5–8], Med-VQA has great potential in the healthcare industry. Med-VQA can support healthcare professionals in their diagnoses and treatment plans, and offer patients informative consultations.

Despite recent advancements in Visual Question Answering (VQA) in the general domain [9–14], Med-VQA is still in its early stages of development and needs lots of efforts. The challenge of Med-VQA lies in the lack of large-scale annotated datasets, particularly ones with external knowledge [15]. Additionally, the medical field involves complex medical terminologies, concepts, and reasoning which

S. Yang and S. Islam (Eds.): APWeb-WAIM 2022 Workshops, CCIS 1784, pp. 3–9, 2023.
https://doi.org/10.1007/978-981-99-1354-1_1

require specialized knowledge to understand. Drawing on combining knowledge for general VQA, we probe the benchmarks and approaches that address the mentioned issues by using extra medical knowledge including knowledge graphs and text information. The knowledge can not only help reinforce the accuracy and reliability of Med-VQA but also improve the reasoning ability of the model [12,16].

In this survey, we aim to provide a comprehensive overview of the state-of-the-art research in knowledge-enhanced Med-VQA, which can advance its development and applications. To accomplish this, we analyze the current Med-VQA datasets that incorporate extra knowledge, examining their characteristics. Moreover, we identify the key technologies used to enhance Med-VQA with knowledge. Finally, we discuss the challenges and limitations faced by the field and provide insights and suggestions for potential solutions.

2   Datasets

As far as our current knowledge goes, there are 10 publicly accessible datasets for Med-VQA research at present including VQA-MED-2018 [17], VQA-RAD [18], VQA-MED-2019 [19], RadVisDial [20], PathVQA [21], VQA-MED-2020 [22], SLAKE [23], VQA-MED-2021 [24], OVQA [25], and P-VQA [15]. They have been elaborated in the published review [6]. In this survey, we specialize in datasets with knowledge graphs like SLAKE [23]and P-VQA [15]. The following paragraphs provide an overview of these two datasets.

2.1   SLAKE

SLAKE [23] is a comprehensive bilingual dataset for Med-VQA with images covering healthy and unhealthy cases with a medical knowledge graph. The images are annotated by experienced physicians and have 642 images with 39 organs and 12 diseases in different modalities (CT, MRI, and X-Ray) and body parts (head, neck, chest, abdomen, and pelvic cavity). Questions were proposed by professional doctors using a flexible annotation system. The knowledge graph was constructed by extracting information from a large-scale knowledge base and refined to cover frequently referenced knowledge. Finally, the dataset includes 642 images with 14,028 question-answer pairs and 5,232 medical knowledge triplets.

2.2   P-VQA

P-VQA [15] points out that current Med-VQA datasets primarily focus on supporting doctors with diagnoses, and may not be accessible or useful for patients without a medical background. The P-VQA dataset aims to improve the medical experience for patients and ease the workload of doctors. It contains images and questions, as well as a knowledge graph built based on common patient questions. A total of 2,169 images come from real hospital cases and use various imaging technologies. The knowledge graph has 419 entities and three types of relationships: disease-to-attribute, disease-to-disease, and attribute-to-attribute. The unidentified questions and answers are generated through templates filled in by trained physicians, with an average question length of 9.62 and 34% of answers coming from multiple entities in the knowledge graph.

2.3 Discussion

The above benchmarks indicate that researchers are beginning to pay attention to the important role of knowledge in Med-VQA, but more sufficient and informative knowledge is needed in the knowledge graph to enhance the performance of Med-VQA. There are several challenges when building datasets with knowledge graphs. Firstly, structuring the medical knowledge and information into a knowledge graph is time-consuming, and choosing the right entities, relations, and attributes is strict to ensure quality. Maintaining the privacy and confidentiality of the patients and ethical considerations should be taken into account as well. Apart from that, datasets with a set of unstructured facts for the questions are also needed to offer useful knowledge for Med-VQA models.

3 Methods

There has been increasing research looking at how to incorporate knowledge in methods for general VQA [12,26–29]. We investigate the knowledge-enhanced approaches used in the Med-VQA task according to the medical characteristics.

Med-VQA approaches mostly adopt the framework [6] widely used in the general domain of VQA, including an image encoder, a question encoder, a multimodal fusion module, and an answer prediction module. A difference is that the current VQA models for natural images often utilize object detectors, scene graph parsers, and action predictors to represent images, which is not available for Med-VQA. Additionally, knowledge-combined approaches usually have an additional encoder for external knowledge representation [15] and the fusion module should manage to deal with features from the three encoders (as shown in Fig. 1). For instance, SLAKE [23] uses VGG16 [30] to encode the image, LSTM [31] to encode the question, and a stacked attention network (SAN) to combine visual and textual features. For the knowledge-based questions, a distance-based method TranE [32] is leveraged to search for the candidate entity in the knowledge graph. Then the fused features of the question and the image are integrated to predict the final answer. After transforming the medical images, questions, and knowledge graph into embeddings, P-VQA [15] develops the fusion module focusing on merging the features from the image-question level, image-disease level, and question-relation level, where the disease and relation are from the knowledge graph. Experiments in these studies show that external knowledge can greatly improve the performance of Med-VQA models.

Besides, some other works [33–36] associate with already existing knowledge bases by retrieving question-related facts or entity linking to build Med-VQA models or datasets. A large structured knowledge base of biomedical terminologies is employed through name entity recognition and entity linking in [34]. The knowledge in the knowledge base is used to align vision and language representations, supplement the knowledge and teach the model to focus on important information. Moreover, there also exists work that constructs the knowledge graph between the questions and the images in the dataset when modeling. Then the knowledge is incorporated into the model to guide the learning.

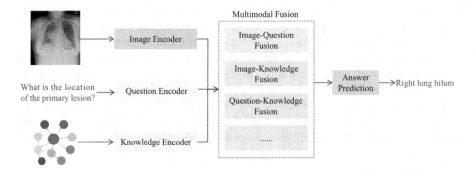

Fig. 1. The common knowledge-enhanced Med-VQA frameworks.

4   Challenge and Future Works

Although knowledge is helpful to boost the performance of Med-VQA and researchers have made some explorations in knowledge-enhanced studies, there are still some challenges. First and foremost, structured medical knowledge is scarce and expensive to obtain since medical data is often sensitive and subject to strict regulations. From the model architecture, integrating external knowledge increases the complexity and needs delicate modification. In addition, the performance of knowledge-enhanced models may depend on the quality and relevance of the knowledge used, and it may be difficult to determine the optimal amount and type of knowledge to use. However, there are some inspirations to alleviate the issues for future research.

Data Sharing and Collaboration. To overcome the scarcity and cost of structured medical knowledge, researchers and organizations need to further share and collaborate on data, enabling the development of larger and more comprehensive medical knowledge bases. Data-sharing initiatives, such as open data platforms and data-sharing agreements, can be used to make structured medical knowledge more accessible and available.

Integrating State-of-Art Technology Paradigms. MEVF [37] adopts model-agnostic meta learning [38] and convolutional denoising autoencoder [39] to deal with the lack of data problem. Moreover, the successful cases in general VQA using contrastive learning, pre-trained models, and knowledge distillation strategies show valuable insights and are worth considering and studying the effectiveness when incorporating knowledge in a semi-supervised framework.

Human-in-the-Loop Systems: To ensure the quality and reliability of the answers provided, human-in-the-loop systems can be used to review and verify the answers generated by the system. Interpretable models and visualization techniques can be explored to understand the impact of different amounts and types of knowledge on performance.

Acknowledgements. The present research was supported by the Fundamental Research Funds for the Central Universities with grant Nos. 22120220069 and the National Natural Science Foundation of China with Grant No. 62176185.

References

1. Devlin, J., Chang, M.W., Lee, K., Toutanova, K.: BERT: pre-training of deep bidirectional transformers for language understanding. In: NAACL-HLT (2019)
2. Chen, F.L., et al.: VLP: a survey on vision-language pre-training. Mach. Intell. Res. **20**(1), 38–56 (2023)
3. Snell, J., Swersky, K., Zemel, R.: Prototypical networks for few-shot learning. In: NIPS (2017)
4. Xian, Y., Lampert, C.H., Schiele, B., Akata, Z.: Zero-shot learning-a comprehensive evaluation of the good, the bad and the ugly. IEEE Trans. Pattern Anal. Mach. Intell. **41**(9), 2251–2265 (2018)
5. Sarrouti, M., Ben Abacha, A., Demner-Fushman, D.: Goal-driven visual question generation from radiology images. Information **12**(8), 334 (2021). https://doi.org/10.3390/info12080334
6. Lin, Z., et al.: Medical visual question answering: a survey. arXiv:2111.10056 (2021)
7. Sengar, N., Joshi, R.C., Dutta, M.K., Burget, R.: EyeDeep-Net: a multi-class diagnosis of retinal diseases using deep neural network. Neural Comput. Appl. 1–21 (2023). https://doi.org/10.1007/s00521-023-08249-x
8. Liu, R., et al.: Application of artificial intelligence-based dual-modality analysis combining fundus photography and optical coherence tomography in diabetic retinopathy screening in a community hospital. Biomed. Eng. Online **21**(1), 1–11 (2022). https://doi.org/10.1186/s12938-022-01018-2
9. Antol, S., et al.: VQA: visual question answering. In: ICCV (2015)
10. Yu, Z., Yu, J., Cui, Y., Tao, D., Tian, Q.: Deep modular co-attention networks for visual question answering. In: CVPR (2019)
11. Zheng, W., Yin, L., Chen, X., Ma, Z., Liu, S., Yang, B.: Knowledge base graph embedding module design for Visual question answering model. Pattern Recogn. **120**, 108153 (2021). https://doi.org/10.1016/j.patcog.2021.108153
12. Marino, K., Rastegari, M., Farhadi, A., Mottaghi, R.: Ok-VQA: a visual question answering benchmark requiring external knowledge. In: CVPR (2019)
13. Ravi, S., Chinchure, A., Sigal, L., Liao, R., Shwartz, V.: VLC-BERT: visual question answering with contextualized commonsense knowledge. In: WACV (2023)

14. Song, L., Li, J., Liu, J., Yang, Y., Shang, X., Sun, M.: Answering knowledge-based visual questions via the exploration of question purpose. Pattern Recogn. **133**, 109015 (2023). https://doi.org/10.1016/j.patcog.2022.109015

15. Huang, J., et al.: Medical knowledge-based network for patient-oriented visual question answering. Inf. Process. Manag. **60**(2), 103241 (2023). https://doi.org/10.1016/j.ipm.2022.103241

16. Chen, Z., Li, G., Wan, X.: Reason and learn: enhancing medical vision-and-language pre-training with knowledge. In: ACM Multimedia (2022)

17. Hasan, S.A., Ling, Y., Farri, O., Liu, J., Müller, H., Lungren, M.P.: Overview of ImageCLEF 2018 medical domain visual question answering task. In: CLEF (Working Notes) (2018)

18. Lau, J.J., Gayen, S., Abacha, A.B., Demner-Fushman, D.: A dataset of clinically generated visual questions and answers about radiology images. Sci. Data **5**, 1–10 (2018). https://doi.org/10.1038/sdata.2018.251

19. Abacha, A.B., Hasan, S.A., Datla, V.V., Liu, J., Demner-Fushman, D., Müller, H.: VQA-med: overview of the medical visual question answering task at ImageCLEF 2019. In: CLEF (working notes) (2019)

20. Kovaleva, O., et al.: Towards visual dialog for radiology. In: Proceedings of the 19th SIGBioMed Workshop on Biomedical Language Processing (2020)

21. He, X., Zhang, Y., Mou, L., Xing, E., Xie, P.: Pathvqa: 30000+ questions for medical visual question answering. arXiv:2003.10286 (2020)

22. Abacha, A.B., Datla, V.V., Hasan, S.A., Demner-Fushman, D., Müller, H.: Overview of the VQA-med task at ImageCLEF 2020: visual question answering and generation in the medical domain. In: CLEF (working notes) (2020)

23. Liu, B., Zhan, L.M., Xu, L., Ma, L., Yang, Y., Wu, X.M.: Slake: a semantically-labeled knowledge-enhanced dataset for medical visual question answering. In: ISBI (2021)

24. Abacha, A.B., et al.: Overview of the VQA-med task at ImageCLEF 2021: visual question answering and generation in the medical domain. In: CLEF (working notes) (2021)

25. Huang, Y., Wang, X., Liu, F., Huang, G.: OVQA: a clinically generated visual question answering dataset. In: SIGIR (2022)

26. Narasimhan, M., Lazebnik, S., Schwing, A.: Out of the box: reasoning with graph convolution nets for factual visual question answering. In: NIPS (2018)

27. Narasimhan, M., Schwing, A.G.: Straight to the facts: learning knowledge base retrieval for factual visual question answering. In: ECCV (2018)

28. Yang, Z., et al.: An empirical study of GPT-3 for few-shot knowledge-based VQA. In: AAAI (2022)

29. Song, L., Li, J., Liu, J., Yang, Y., Shang, X., Sun, M.: Answering knowledge-based visual questions via the exploration of question purpose. Pattern Recogn. **133**, 109015 (2023)

30. Simonyan, K., Zisserman, A.: Very deep convolutional networks for large-scale image recognition. In: ICLR (2015)

31. Hochreiter, S., Schmidhuber, J.: Long short-term memory. Neural Comput. **9**(8), 1735–1780 (1997)

32. Bordes, A., Usunier, N., Garcia-Duran, A., Weston, J., Yakhnenko, O.: Translating embeddings for modeling multi-relational data. In: NIPS (2013)

33. Luo, F., Zhang, Y., Wang, X.: IMAS++: an intelligent medical analysis system enhanced with deep graph neural networks. In: CIKM (2021)

34. Chen, Z., Li, G., Wan, X.: Align, reason and learn: enhancing medical vision-and-language pre-training with knowledge. In: ACMMM (2021)

35. Zheng, W., Yan, L., Wang, F.Y., Gou, C.: Learning from the guidance: knowledge embedded meta-learning for medical visual question answering. In: ICONIP (2020)
36. Huang, Y., Wang, X., Liu, F., Huang, G.: OVQA: a clinically generated visual question answering dataset. In: SIGIR (2022)
37. Nguyen, B.D., Do, T.T., Nguyen, B.X., Do, T., Tjiputra, E., Tran, Q.D.: Overcoming data limitation in medical visual question answering. In: MICCAI (2019)
38. Finn, C., Abbeel, P., Levine, S.: Model-agnostic meta-learning for fast adaptation of deep networks. In: ICML (2017)
39. Masci, J., Meier, U., Cireşan, D., Schmidhuber, J.: Stacked convolutional autoencoders for hierarchical feature extraction. In: ICANN (2011)

Effectively Filtering Images for Better Multi-modal Knowledge Graph

Huang Peng, Hao Xu$^{(\boxtimes)}$, Jiuyang Tang, Jibing Wu, and Hongbin Huang

Laboratory for Big Data and Decision, National University of Defense Technology, Changsha 410000, China
{phmail,xuhao}@nudt.edu.cn

Abstract. The existing multi-modal knowledge graph construction techniques have become mature for processing text modal data, but lack effective processing methods for other modal data such as visual modal data. Therefore, the focus of multi-modal knowledge graph construction lies in image and image and text fusion processing. At present, the construction of multi-modal knowledge graph often does not filter the image quality, and there are noises and similar repetitive images in the image set. To solve this problem, this paper studies the quality control and screening of images in the construction process of multi-modal knowledge graph, and proposes an image refining framework of multi-modal knowledge graph, which is divided into three modules. The final experiment proves that this framework can provide higher quality images for multi-modal knowledge graphs, and in the benchmark task of multi-modal entity alignment, the effect of entity alignment based on the multi-modal knowledge graphs constructed in this paper has been improved compared with previous models.

Keywords: Multi-modal knowledge graph · Clustering algorithm · Image-text matching · Entity alignment

1 Introduction

With the continuous development of information technology in Internet era, web technology is evolving from web links to data links, and gradually to the semantic web [2]. On this basis, Google formally proposed the concept of knowledge graph, which is the sublimation of semantic web technology and standard. At present, with the continuous development of intelligent information services, knowledge graph technology has been widely used in intelligent search, intelligent question and answer, personalized recommendation and other fields.

With the development of computer vision and multi-modal learning, more and more algorithm and models are proposed to extract visual data feature information, which provides technical support for the development of traditional knowledge graph to multi-modal knowledge graph. In addition, studies have found that adding visual modal information to knowledge graph can provide

S. Yang and S. Islam (Eds.): APWeb-WAIM 2022 Workshops, CCIS 1784, pp. 10–22, 2023.
https://doi.org/10.1007/978-981-99-1354-1_2

important help for knowledge graph completion and triplet classification [2,17], and also improve the reasoning ability of knowledge graph [10]. Therefore, multi-modal knowledge graph, especially knowledge graph with visual modal information, has gradually become a research hotspot.

Without losing generality, we only consider knowledge graph with visual modal and text modal. The current construction of multi-modal knowledge graph that contains visual and text modal is mostly based on the existing traditional text knowledge graph, add the entity related images as visual modal entity, and set up cross modal semantic relations between different modal entities, so that images' quality becomes an important factor in evaluating the quality of multi-modal knowledge graph construction. Most of the existing methods use the Internet image search engine to search and obtain related images with the entity name as the key word. However, the current main methods generally have the following two problems [18]:

(1) The query results returned by image search engines have obvious image duplication or high similarity, noisy data, and inaccurate images caused by polysemy. Most of the existing methods do not filter these images at all, and few only utilizes simple algorithm with low efficiency, and leads to a low quality graph.
(2) The information other than the image is not fully utilized to help filter the image. It is difficult to judge whether the image and text have the same semantic meaning only by using image information. However, there are accurate text descriptions of entities in most encyclopedias, with which the entities with polysemy can be distinguished and refined.

Therefore, this paper proposes an image refining framework (IRF), which can obtain highly relevant and diverse images for entities in the knowledge graph. The construct process of multi-modal knowledge graph consists of three modules: The image data acquisition module uses the Internet image search engine to crawl the corresponding image of the entity, and preliminarily filters the repeated or similar images based on clustering algorithm. The multi-modal fusion filtering module embedded the image and text into a same semantic space based on image-text matching model and filter the images with low relevance. In addition, we introduce multi-modal entity alignment to verify the validity of our image filtering model, which helps generate better entities' visual representation.

The main contributions of the framework proposed in this paper can be summarized as follows:

– The image acquisition module is designed to conduct preliminary filtering of unsupervised image data based on clustering algorithm, partly solving the problem of high repetition or similarity of images obtained by Internet image search engine.
– A multi-modal fusion filtering module is designed. The text description information of entities in encyclopedia is used to filter irrelevant image entities based on the pretrained text matching model, so that entities can obtain higher quality and diversified images and improve the quality of multi-modal knowledge graph visual modal data.

– Leverage high quality image after filtering to generate more accurate visual feature representation of entities, and achieve better results than existing methods in multi-modal entity alignment tasks.

2 Related Work

2.1 Multi-modal Knowledge Graph

At present, most multimodal knowledge graphs are knowledge graphs with entity images, among which IMGpedia [4], MMKG [9] and Richpedia [15] are representative. IMGpedia collects a large amount of visual information from the Wikimedia Commons dataset, while also linking with DBpedia [7] and DBpedia Commons to provide semantic context and further metadata. However, there are some defects such as sparse relationship types, small number of relationships, and unclear image classification. MMKG is a multi-modal knowledge graph jointly constructed by the National University of Singapore and Xiamen University. Its entities are extracted from FreeBase, DBpedia and YAGO, while the images of entities are obtained from Google, Bing and Yahoo image search engines. MMKG is specially built for the completion of text knowledge graphs, mainly for small datasets, and also does not consider the diversity of images. In 2019, Wang Meng's team from Southeast University added sufficient and diverse images to the textual entities in Wikidata [13] through network links and image search, and constructed a more comprehensive multimodal knowledge graph Richpedia. Richpedia filters images through a series of preprocessing operations, including denoising operations and diversity detection, to ensure the relevance and diversity of image entities. However, it only uses a clustering algorithm without external information, resulting in poor filtering effect.

2.2 Multi-modal Entity Alignment

The aim of multi-modal entity alignment is to recognize the same multi-modal entity from different multi-modal graph. Different from entity alignment, multi-modal entity alignment requires utilizing and fusing information from multiple modalities. Multimodal entity alignment is a relatively new direction, and there are few researches directly oriented to this task. Image-embodied Knowledge Representation Model (IKRL) [17] learns knowledge representation through triples and images. At first, a neural image encoder is used to construct representation for all images of entities. These image representations are then aggregated into an integrated image-based representation of entities by an attention-based approach. Literature [9] uses Product of Experiment (PoE) model to find potential aligned entities by integrating similarity scores of structures, attributes and visual features. The literature [16] proposed GCN-align, which uses graph convolutional neural networks to generate structural and visual features of entities, and combines them with fixed weights for entity alignment.

3   Model

The current multi-modal knowledge graph generally has the disadvantage of low image quality, that is, there are low correlation and high repetition images. To solve this problem, this paper designs an Image Refining Framework (IRF) based on the Richpedia construction process [15] to obtain high-quality entity images through a series of operations. The framework process is divided into three modules: image data acquisition module, multi-modal fusion and filtering module, multi-modal entity alignment module.

3.1   Image Acquisition Module

The current visual modal data of multi-modal knowledge graph mainly comes from Internet search engine, thus there inevitably exists repetitive or similar images, which will decrease the proportion of valid visual information in the final graph without filtering. As shown in Fig. 1, when Saving Private Ryan is searched in the search engine directly, there are several similar pictures returned, some of which are movie posters and some are stills with different sizes and hues. Therefore, this module will use the image search engine to obtain images of entities, and then carry out preliminary de-duplication and de-noising based on the clustering algorithm.

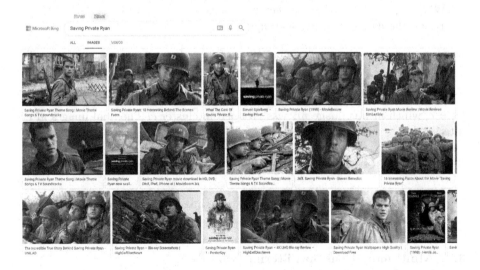

Fig. 1. Returned result of "Saving Private Ryan" by Bing

Firstly, the entity name is used as the key word to retrieve the entity related images in the image search engine, and the crawler program is written to automatically obtain a certain number of images. After a sufficient number of images are crawled, a ResNet [5] model pre-trained on ImageNet [3] is used to convert them into a one-dimensional feature vector of 2048 length, and then performs clustering using DBSCAN [1] with Euclidean distance as the similarity metric. After clustering, one image is selected from each cluster as representative, which completes a preliminary filtering of the image (Fig. 2).

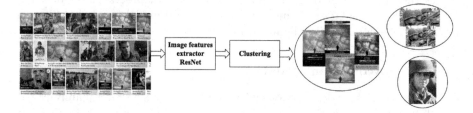

Fig. 2. Filtering procedure based on clustering

Set images of entity i crawled from image search engine as $S_i = \{m_1^i, m_2^i, \ldots, m_N^i\}$, where N is the quantity of images for each entity. Using ResNet to extract image features:

$$f_j^i = \texttt{ResNet}(m_j^i) \tag{1}$$

Where $f_j^i \in R^{2048}$ is feature vector of image m_j^i. Then utilizing DBSCAN to cluster all these images and get grouped images:

$$C^i = \texttt{DBSCAN}(f_1^i, f_2^i, \ldots, f_N^i) \tag{2}$$

Where $C^i \in R^N$, and if image m_j^i was judged as noise point, $C_j^i = -1$, else $C_j^i = c$, which means image m_j^i was grouped in cluster c. The final images after filtering are:

$$S_i' = \{m_j^i \in S_i | (C_j^i = -1) \vee (\forall m_k^i \in S_i (k \neq j \wedge C_k^i = C_j^i \rightarrow m_k^i \notin S_i'))\} \tag{3}$$

S_i' is set of images after filtering, containing images that judged as noise point or one of each cluster.

3.2   Multi-modal Fusion Filtering Module

After the preliminary filtering in last section, there are still images that relevant less to entity, or images of other entities because of ambiguity. In this case, additional information is needed to help judge and filter the images. Therefore, in the absence of labeled data, this module adopts the method of extracting the entity text description from Wikidata [13], using the pre-trained image-text matching model CVSE [14] to generate embeddings of images and text, and calculating their similarity; Then, by setting the similarity threshold to filter the relevant images of the entity and drop the low-correlation noise images.

Firstly, the corresponding entries of the entity are searched from Wikidata, and the description text about the entity in this entry is crawled, and then the image and text data are fed into the pre-trained CVSE model (Fig. 3).

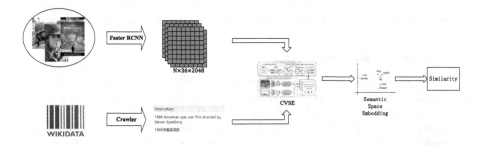

Fig. 3. Image filter process

This module uses pre-trained CVSE to calculate the embeddings of image and text and calculates the image-text similarity score. The input of the CVSE model are raw text and the embedding of images, so according to the requirements of the model, the embedding of the images $p_i \in R^{n \times 36 \times 2048}$ is required, where n is the number of images in the image set corresponding to entity i. In order to generate image embedding, this module uses the object detection model Faster R-CNN [12] to generate a feature matrix of 36×2048 for each image by controlling the number of its detection regions, so as to obtain the image set embedding p_i. Next, the image-set embedding p_i and raw text description l_i of entity i are used as the input of the CVSE model to obtain the consensus level embedding of image-set embedding and text description:

$$(\mathbf{v}_i, \mathbf{t}_i) = \text{CVSE}(p_i, l_i) \tag{4}$$

Where $\mathbf{v}_i \in R^{n \times 1024}, \mathbf{t}_i \in R^{1024}$. Then calculate similarity of text and images utilizing cosine measure:

$$
\begin{aligned}
score_i &= (\cos(\mathbf{t}_i, \mathbf{v}_{i1}), \cos(\mathbf{t}_i, \mathbf{v}_{i2}), \ldots, \cos(\mathbf{t}_i, \mathbf{v}_{in})) \\
&= (\frac{\mathbf{t}_i \mathbf{v}_{i1}^{\text{T}}}{|\mathbf{t}_i||\mathbf{v}_{i1}|}, \frac{\mathbf{t}_i \mathbf{v}_{i2}^{\text{T}}}{|\mathbf{t}_i||\mathbf{v}_{i2}|}, \ldots, \frac{\mathbf{t}_i \mathbf{v}_{in}^{\text{T}}}{|\mathbf{t}_i||\mathbf{v}_{in}|}) \\
&= \frac{\mathbf{t}_i}{|\mathbf{t}_i|} (\frac{\mathbf{v}_{i1}^{\text{T}}}{|\mathbf{v}_{i1}|}, \frac{\mathbf{v}_{i2}^{\text{T}}}{|\mathbf{v}_{i2}|}, \ldots, \frac{\mathbf{v}_{in}^{\text{T}}}{|\mathbf{v}_{in}|})
\end{aligned} \tag{5}
$$

In this formula, $\mathbf{v}_{ij}$ represents the consensus level embedding of j-th image in image set corresponding to entity i, namely the j-th row vector of $\mathbf{v}_i$. $\mathbf{v}_{ij}^{\text{T}} \in$

$R^{1024 \times 1}$ a column vector is transpose of $\mathbf{v}_{ij}$. Set threshold of similarity as α, then the image set after filter will be:

$$ims_i' = \{j \mid j \in ims_i, score_{ij} > \alpha\} \tag{6}$$

Where ims_i is serial number set of images before filtering, and ims_i' is serial number set of images after filtering. $score_{ij}$ is j-th component of $score_i$, which is similarity of text description l_i and image m_j corresponding to entity i.

3.3   Multi-modal Alignment Module

After the filtering of entity images, this module will use the visual feature information of the image set corresponding to the entity to perform multi-modal entity alignment, so as to verify the quality of the filtered image set.

Firstly, for the filtered image set ims_i', the similarity between the image and the entity description text is used as the weight after softmax normalization to generate a more accurate visual feature representation of the entity:

$$v_i = \texttt{softmax}(score_i')F_i' \tag{7}$$

Where $v_i \in R^{2048}$ represents the weighting visual feature of entity i, $score_i' \in R^{n'}$ represents the similarity of filtered image set ims_i', while n' is the quantity of images after filtering. $F_i' = \left[f_1^{i}{}^{\mathrm{T}} \; f_2^{i}{}^{\mathrm{T}} \cdots f_{n'}^{i}{}^{\mathrm{T}} \right] \in R^{n' \times 2048}$ is the feature matrix composed of feature vectors $f^i \in R^{2048}$, which generated by ResNet.

As an aspect of graph feature information, structural features are often considered for entity alignment, but their effectiveness in different graph will affect the alignment effect, especially for entities with lack of structural information, the contribution of structural features to the alignment is very limited. But the effectiveness of the visual features is not affected by such, because the visual characteristics of the construction is based on the correlation of external information. This is an advantage of entity alignment based on visual features.

For two multi-modal knowledge graphs $MG_1 =< E_1, R_1, T_1, I_1 >, MG_2 =< E_2, R_2, T_2, I_2 >$, where E, R, T, I represent entities, relations, triples and images respectively, and I_2 is the image set obtained by the image acquisition module and the multi-modal fusion filtering module. Denote $E_1 \in E_1, e_2 \in E_2$ as two entities in two graphs, and their corresponding image sets are $S_1 = \{m_1^1, m_2^1, \ldots, m_{n_1}^1\}, S_2 = \{m_1^2, m_2^2, \ldots, m_{n_2}^2\}, n_1, n_2$ are the corresponding number of images of the two entities respectively. Firstly, Eq. 7 is applied to convert the images in the two image sets into 2048 length one-dimensional feature vectors, and two feature matrices $F_1 = \left[f_1^{1}{}^{\mathrm{T}} \; f_2^{1}{}^{\mathrm{T}} \cdots f_{n_1}^{1}{}^{\mathrm{T}} \right]^{\mathrm{T}} \in R^{n_1 \times 2048}, F_2 = \left[f_1^{2}{}^{\mathrm{T}} \; f_2^{2}{}^{\mathrm{T}} \cdots f_{n_1}^{2}{}^{\mathrm{T}} \right]^{\mathrm{T}} \in R^{n_2 \times 2048}$. For F_1, the mean of its row vector is directly calculated as its visual feature:

$$v_1 = \frac{1}{n_1} \sum_{i=1}^{n_1} f_i^1 \tag{8}$$

The visual feature V_2 is obtained by applying Eq. 7 to F_2, where $v_1, v_2 \in R^{2048}$. Cosine similarity is still used as the similarity measure of visual feature vectors, then:

$$sim(e_1, e_2) = sim(v_1, v_2) = \frac{v_1 v_2^T}{|v_1||v_2|} \tag{9}$$

The distance between the matched entity pairs should be small and the similarity between them should be large. Therefore, for each entity in MG_2, the similarity of all the entities in MG_1 should be calculated, and the candidate matching entities should be returned by sorting the similarity from the largest to the smallest.

4   Experiments

4.1   Experiment Settings

Environment and Parameters. The experimental environment in this section is 64-bit Ubuntu 16.04 LTS, the processor is 8-core Intel® CoreTM I7-7700K@4.20 GHz, the memory size is 62.8GiB, the graphics processor is GeForce GTX 1080. The programs used in the experiments were written in Python 3.6. The number of pictures N crawled for each entity in the image data acquisition module is 50, and the parameters of the clustering algorithm DBSCAN are set to eps=12 and min_samples=2. The image set of each entity in the multi-modal fusion filtering module is obtained after preliminary filtering by clustering in the image data acquisition module. The ResNet used in this paper is a deep 152-layer version with parameters pre-trained on ImageNet. Faster R-CNN uses ResNet101 as the backbone network pre-trained parameters on Visual Genome [6]. CVSE uses parameters pre-trained on the MSCOCO [8] and Flickr30k [11] datasets.

Data Source and Evaluating Metrics. In the image data acquisition module and multi-modal fusion filtering module, the source of images is the query results of image search engines (Bing, Baidu, Sogou), and the source of text description of entities is the description field of the corresponding entry page of Wikidata. In multi-modal entity alignment module in this article the MMKG [9] is uesd as the multi-modal data sets, MMKG data sets, including FB15K, DB15K and YAGO15K, are extracted from FreeBase, DBpedia [7] and Yago.

In the experiment of entity image collection, the quality of collected images is evaluated by the expert scoring method. In the multi-modal entity alignment experiment, Hits@k and Mean Reciprocal Rank (MRR) were used as evaluation metrics. In the following results table, the results of Hits@k will be expressed as percentages, and the best results under the same evaluation metric will be marked in bold.

4.2   Entity Image Collection Experiment

This experiment aimed to collect relative images for entity utilizing image acquisition module and multi-modal fusion filtering module, and demonstrate IRF's
advantage by comparing with existing multi-modal graph.

In this experiment, 3819 entities' descriptions were crawled from Yago website. Then, three existing multi-modal knowledge graphs IMGpedia, Richpedia
and MMKG are selected to compare the image quality. Since there is no quantitative metrics of image quality, this experiment uses experts scoring method
to formulate three metrics: accuracy, diversity and quantity, and the priority
is accuracy > diversity > quantity. Only when the higher priority metric is
consistent or similar, the lower priority metric will be compared. The set of
comments is better, fairly, worse. In this experiment, 50 entities are randomly
selected from the intersection of three multimodal knowledge graphs, and 12
experts are invited to evaluate and compare the image sets of these entities in
the three graphs with the experimental results. The final evaluation results are
in the form of the percentage of entities in each comment given by experts. The
final evaluation results are shown in Fig. 4. The quality of the images obtained
in this experiment is significantly better than that of the images in MMKG and
IMGpedia. Compared with Richpedia, there are still 46% of the entities have
images with better quality than Richpedia.

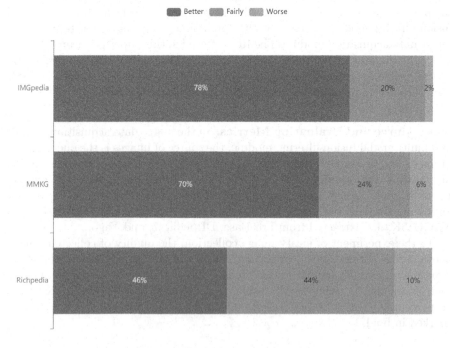

Fig. 4. Results of expert scoring

4.3   Multi-modal Entity Alignment Experiment

To prove the effectiveness of IRF's mapping, this section sets up a multimodal entity alignment experiment and compares the results with three existing methods IKRL, GCN-Align and PoE.

The datasets used in the experiments are FB15K-DB15K and FB15K-YAGO15K. Based on the experimental setup in the literature, the proportion of seed entities was set to 20% and 50% in this experiment. Note that all the models used in this framework are pre-trained, so there is no training process; However, the results are different under the condition of different proportion of seed entities, the reason is that when the proportion of seed entities is large, the number of entities for verification is small, so the pre-aligned entities are easy to get a higher ranking, and each index also improves. For IKRL and GCN-Align, this experiment uses their open source codes to reproduce and test. However, the source code of PoE model is not given in the original paper, so the results given in the original paper are directly used for comparison. The alignment results are shown in Table 1.

According to Table 1, it is obvious that the IRF designed in this paper achieves better results than IKRL, GCN-Align and PoE in all indicators on the two datasets. Especially, when the proportion of seed entities on the FB15K-YAGO15K dataset is 50%, the Hits@1 index is more than 20% points higher than the best PoE.

Table 1. Entity alignment results

Dataset	Method	Seed=20%			Seed=50%		
		Hits@1(%)	Hits@10(%)	MRR	Hits@1(%)	Hits@10(%)	MRR
FB15K-DB15K	IKRL	2.96	11.45	0.059	5.53	24.14	0.121
	GCN-align	6.26	18.81	0.105	13.79	34.6	0.21
	PoE	11.1	17.8	-	23.5	33	-
	IRF	**17.75**	**34.14**	**0.233**	**29.62**	**50.45**	**0.368**
FB15K-YAGO15K	IKRL	3.84	12.50	0.075	6.16	20.45	0.111
	GCN-align	6.44	18.72	0.106	14.09	34.8	0.209
	PoE	8.7	13.3	-	18.5	24.7	-
	IRF	**21.65**	**40.22**	**0.282**	**35.72**	**55.83**	**0.419**

4.4   Ablation Experiment

The main parts of the framework that play a major role in filtering images are the clustering algorithm in the image acquisition module and the image-text matching model in the multi-modal fusion filtering module. To confirm the effectiveness of the two parts for the final multi-modal entity alignment task, an ablation experiment is designed for comparative verification in this section. IRF$_{-Cluster}$ and IRF$_{-Fusion}$ represent the IRF after removing the clustering algorithm and the whole second model, respectively.

Specifically, IRF$_{-\text{Cluster}}$ directly sends the images from the image search engine to the multi-modal fusion screening module. IRF$_{-\text{Fusion}}$ directly embeds the image set obtained by the image acquisition module with ResNet without using text description. The effectiveness of each module is tested by comparing with the complete model.

This experiment was conducted on FB15K-DB15K and FB15K-YAGO15K datasets, and the proportion of seed entities was set to 20% and 50% for comparison. The results are shown in Tables 2. It can be seen from the results in the table that the full model obtains the best effect in all cases, and the performance of the model results after ablation decreases. IRF$_{-\text{Fusion}}$ fells the most obviously, and IRF$_{-\text{Cluster}}$ with the removal of clustering algorithm decreases relatively little, but it also decreases by 3.6% points under the condition of 50% seed entity proportion on FB15K-YAGO15K. The experimental results show that the CVSE model has the greatest impact on entity alignment, and it is confirmed that the fusion of text modality and visual modality information can significantly improve the effect of entity alignment. In addition, the experiment also finds that although the performance of IRF$_{-\text{Cluster}}$ is relatively small, the images of IRF$_{-\text{Cluster}}$ are duplicated or similar, so the clustering algorithm mainly plays an important role in image quality control.

Table 2. Entity alignment results

Dataset	Method	Seed=20%			Seed=50%		
		Hits@1(%)	Hits@10(%)	MRR	Hits@1(%)	Hits@10(%)	MRR
FB15K-DB15K	IRF$_{-\text{Cluster}}$	16.19	31.01	0.212	26.29	48.66	0.360
	IRF$_{-\text{Fusion}}$	14.13	28.77	0.191	22.91	43.08	0.297
	IRF	**17.75**	**34.14**	**0.233**	**29.62**	**50.45**	**0.368**
FB15K-YAGO15K	IRF$_{-\text{Cluster}}$	19.75	37.38	0.255	32.08	53.24	0.393
	IRF$_{-\text{Fusion}}$	15.84	32.36	0.216	27.38	48.14	0.345
	IRF	**21.65**	**40.22**	**0.282**	**35.72**	**55.83**	**0.419**

5 Conclusions

Existing multimodal knowledge graphs such as MMKG and Richpedia build multi-modal knowledge graphs by adding image entities and corresponding multi-modal relationships, but often do not effectively filter the quality of images, and there are noise and similar duplicate images in the image set. To solve this problem, this paper studies the quality filtering and processing of images in the construction process of multi-modal knowledge graph, designs a multi-modal knowledge graph image refinement framework. The entity image collection exper-iment proves that the framework can provide higher quality images for multimodal knowledge graphs. The multi-modal entity alignment experiment shows that the alignment effect of the framework is improved compared with the previous models on the benchmark task of multi-modal entity alignment.

References

1. Anant, R., Sunita, J., Jalal, A.S., Manoj, K.: A density based algorithm for discovering density varied clusters in large spatial databases. Int. J. Comput. Appl. **3**(6) (2010)
2. Bizer, C., Heath, T., Berners-Lee, T.: Linked data: the story so far. Int. J. Semant. Web Inf. Syst. **5**, 1–22 (2009). https://doi.org/10.4018/jswis.2009081901
3. Deng, J., Dong, W., Socher, R., Li, L.J., Li, K., Fei-Fei, L.: Imagenet: a large-scale hierarchical image database. In: 2009 IEEE Conference on Computer Vision and Pattern Recognition, pp. 248–255. IEEE (2009)
4. Ferrada, S., Bustos, B., Hogan, A.: IMGpedia: a linked dataset with content-based analysis of wikimedia images. In: d'Amato, C., et al. (eds.) ISWC 2017. LNCS, vol. 10588, pp. 84–93. Springer, Cham (2017). https://doi.org/10.1007/978-3-319-68204-4_8
5. He, K., Zhang, X., Ren, S., Sun, J.: Deep residual learning for image recognition. In: Proceedings of the IEEE Conference on Computer Vision and Pattern Recognition, pp. 770–778 (2016)
6. Krishna, R., et al.: Visual genome: connecting language and vision using crowd-sourced dense image annotations. Int. J. Comput. Vision **123**(1), 32–73 (2017)
7. Lehmann, J., et al.: Dbpedia-a large-scale, multilingual knowledge base extracted from wikipedia. Semant. Web **6**(2), 167–195 (2015)
8. Lin, T.-Y., et al.: Microsoft COCO: common objects in context. In: Fleet, D., Pajdla, T., Schiele, B., Tuytelaars, T. (eds.) ECCV 2014. LNCS, vol. 8693, pp. 740–755. Springer, Cham (2014). https://doi.org/10.1007/978-3-319-10602-1_48
9. Liu, Y., Li, H., Garcia-Duran, A., Niepert, M., Onoro-Rubio, D., Rosenblum, D.S.: MMKG: multi-modal knowledge graphs. In: Hitzler, P., et al. (eds.) ESWC 2019. LNCS, vol. 11503, pp. 459–474. Springer, Cham (2019). https://doi.org/10.1007/978-3-030-21348-0_30
10. Mousselly-Sergieh, H., Botschen, T., Gurevych, I., Roth, S.: A multimodal translation-based approach for knowledge graph representation learning. In: Proceedings of the Seventh Joint Conference on Lexical and Computational Semantics, pp. 225–234. Association for Computational Linguistics (2018). https://doi.org/10.18653/v1/S18-2027
11. Plummer, B.A., Wang, L., Cervantes, C.M., Caicedo, J.C., Hockenmaier, J., Lazebnik, S.: Flickr30k entities: collecting region-to-phrase correspondences for richer image-to-sentence models. In: Proceedings of the IEEE International Conference on Computer Vision, pp. 2641–2649 (2015)
12. Ren, S., He, K., Girshick, R., Sun, J.: Faster R-CNN: towards real-time object detection with region proposal networks. In: Advances in Neural Information Processing Systems, vol. 28 (2015)
13. Vrandečić, D., Krötzsch, M.: Wikidata: a free collaborative knowledgebase. Commun. ACM **57**(10), 78–85 (2014)
14. Wang, H., Zhang, Y., Ji, Z., Pang, Y., Ma, L.: Consensus-aware visual-semantic embedding for image-text matching. In: Vedaldi, A., Bischof, H., Brox, T., Frahm, J.-M. (eds.) ECCV 2020. LNCS, vol. 12369, pp. 18–34. Springer, Cham (2020). https://doi.org/10.1007/978-3-030-58586-0_2
15. Wang, M., Qi, G., Wang, H.F., Zheng, Q.: Richpedia: a comprehensive multi-modal knowledge graph. In: Wang, X., Lisi, F.A., Xiao, G., Botoeva, E. (eds.) JIST 2019. LNCS, vol. 12032, pp. 130–145. Springer, Cham (2020). https://doi.org/10.1007/978-3-030-41407-8_9

16. Wang, Z., Lv, Q., Lan, X., Zhang, Y.: Cross-lingual knowledge graph alignment via graph convolutional networks. In: Proceedings of the 2018 Conference on Empirical Methods in Natural Language Processing, pp. 349–357 (2018)
17. Xie, R., Liu, Z., Luan, H., Sun, M.: Image-embodied knowledge representation learning. In: IJCAI, pp. 3140–3146 (2017). https://doi.org/10.24963/ijcai.2017/438
18. Xueyao, J., Weichen, L., Jingping, L., Zhixu, L., Yanghua, X.: Entity image collection based on multi-modality pattern transfer(). Comput. Eng. **48**(08) (2022). https://doi.org/10.19678/j.issn.1000-3428.0064039

Accelerating Distributed Knowledge Graph System Based on Non-Volatile Memory

Yuhang Li[1], Weiming Li[1], Yangyang Wang[1], Yanfeng Chai[1,2(✉)], and Yunpeng Chai[1(✉)]

[1] Key Laboratory of Data Engineering and Knowledge Engineering, MOE, and School of Information, Renmin University of China, Beijing, China
ypchai@ruc.edu.cn
[2] College of Computer Science and Technology, Taiyuan University of Science and Technology, Taiyuan, China
yfchai@tyust.edu.cn

Abstract. Deploying large scale knowledge graphs on distributed systems has become an industry trend for their high scalability and availability. There are some distributed graph databases that prefer to adopt the NoSQL data models like the key-value store as their storage engines for its scalability and practicability. Therefore, an upper-level graph query language (GQL) statement will be translated into a group of the native and hybrid kinds of key-value (KV) operations. To accelerate the KV operations generated form upper-level knowledge graph queries, we propose a high performance knowledge graph system with a non-volatile memory based queries booster (KGB). KGB mainly contains a neighbors queries auxiliary index for reducing KVs searching cost, a fast Raft algorithm for efficient KVs operations, and a KV tuning mechanism to acquire extra performance promotion for knowledge graph application scenarios. Experiments show that KGB can effectively reduce the latency and achieve higher performance promotion for knowledge graph system.

Keywords: Knowledge graph · Key-value stores · Non-volatile memory (NVM) · performance optimization

1 Introduction

Knowledge graph networks can effectively express the complex relationships between different types of things, and have a wide range of applications in different fields of real life. With the rise and development of knowledge graph and graph computing, the processing and analysis of large-scale graph data have gradually become a hot issue in the industry. Although graph computing has shown its advantages and great development prospects in various neighborhoods, it has also brought great challenges to the platform that provides underlying support and storage for related data. We design and implement a high performance knowledge graph system based NVM, and optimize our system in the following three aspects:

S. Yang and S. Islam (Eds.): APWeb-WAIM 2022 Workshops, CCIS 1784, pp. 23–33, 2023.
https://doi.org/10.1007/978-981-99-1354-1_3

(1) Neighbors queries auxiliary index. Implement neighbors index and stores in NVM to speed up the query performance; Query the results in index instead of repeating traversal graph, to decrease the interaction between storage interface and consensus layer.
(2) A fast Raft algorithm for efficient KVs operations. When raft replicates data to more than half of the nodes, we process the data apply and the response client in parallel. This greatly improves the efficiency of writing data.
(3) A KV tuning mechanism. We tune RocksDB parameters to acquire extra performance promotion for knowledge graph application scenarios.

2   Related Work

RDF or Property Graphs. RDF (Resource Description Framework) is one of the popular knowledge representation forms recommended by W3C, which uses a series of *subject-predicate-object* (SPO) triples to describe the relationship. Property Graph (PG) is more popular among the NoSQL systems, which can be easily implemented as the key-value model for distributed systems. In PG, the data is organized as nodes, relationships, and properties. A node is an entity that can have zero or more properties, which are represented as key-value pairs. Relationships link two nodes in a directed way, which also contain zero or more properties.

Large Scale Graph Storage. Neo4j [16] is a traditional and famous graph database, which is a little strenuous to deal with big data under distributed environment. Trident [20] propose an adaptive storage architecture with the independent layouts of binary tables for different node- and edge-centric access patterns on a centralized system, which can load and process KGs with up to 100 billion edges with lower hardware cost. Columnar graph storage [11] learned some techniques from columnar RDBMSs, which does well in read-heavy analytical workload and data compression. OntoSP [14] proposes a semantic-aware partition method based on ontology hierarchy for RDF graphs in the distributed environment. It can utilize the hierarchical structure of ontology in RDF graphs as heuristic information to promote partitioning performance.

Distributed Key-Value Store. Traditional relational databases normally have some compatibility burdens and no open-source version, which significantly impede the motivations to improve the architecture for performance promotion. With the key-value stores drawing more and more attention, traditional relational database systems are trying to utilize it as the storage engine, e.g., RocksDB [17] in MyRocks [15]. Key-value stores are widely used in distributed database system as low-level storage engine such as TiDB, CockroachDB, FoundationDB in Snowflake [7], and TDSQL. The in-memory graph database A1 [3] utilizes the combination of local DRAM and RDMA to build a distributed graph database with the key-value data model.

For distributed graph database, key-value stores have already become the fundamental storage techniques for their high scalability. NebulaGraph, HugeGraph, and ArangoDB offer options for users to utilize the key-value store

RocksDB as their storage layer. Dgraph [12] also adopts a key-value store Badger [1] as its storage engine. For instance, the read and write operations ratio in Facebook has reached 2:1, i.e., 33.3% write percentage [4,9]. Therefore, database designers are more willing to chose Log Structure Merge Trees (LSM-trees) key-value store for its better space and write amplification.

vRaft [21] modifies the distributed consensus protocol Raft with fast return and follower read mechanism, which could solve the leader node's performance degeneration caused by the neighbors in the virtual cloud environment. KV-Raft [22] further proposes the accessible Raft log mechanism of the distributed consensus protocol with the commit return and the immediate read to accelerate the read/write procedures without breaking the linearizability constraints [10] under the distributed key-value store.

Adaptive & Auto-Tuning. This tendency requires much more knowledge about the database's architecture. Therefore, more related works focus on the open-source database based on key-value stores. SILK [2] focuses on the latency spikes in LSM-Tree and designs an I/O scheduler to dynamic allocate I/O resources according to the priorities of internal operations. CruiseDB focus on the service level agreement (SLA) like throughput or tail-latency, and reconstruct LSM-tree's architecture to have independent right to control the I/O schedule. LDC [5] and ALDC [6] also focus on LSM-tree's merging operation with controllable granularity Low-level Driven Compaction (LDC) method to tune the performance for users specified demands like better tail latency.

Self-adaptive database has been becoming a brand new hot field in recent years. AC-Key [23] subdivides the cache module into the key-value cache, key-pointer cache, and block cache. AC-Key adopts self-adaptive methods to adjust the three's sizes according to the different workload with the Adaptive Replacement Cache (ARC) algorithm to improve caching efficiency. Lethe [19] pays more attention to the out-of-place deletion of the LSM-tree key-value store, while introducing a Fast Deletion (FADE) method to guarantee the delete persistence latency without hurting the performance and resources. CuttleTree [18] designed an adaptive LSM-tree based on the run-time statistics of workload patterns to control the shape of LSM-tree for performance auto-tuning. Monkey [8] focuses on the LSM-tree's bloom filter and other parameters' co-tuning optimization for performance acceleration.

3   Design of Knowledge Graph Booster

Like most graph database systems, our system has three parts: query layer, storage layer, and meta layer. Our Knowledge Graph Booster is all implemented in storage layer as Fig. 1 shows. In the storage layer, the storage interface processes query from query layer. Queries about neighborhoods can be sped up by index in NVM to improve part of reading performance. Then a commit return method in Raft is implemented in consensus layer, which can improve write performance in distributed system. Furthermore, tuning some options in RocksDB store engine also promotes reading and writing performances. We will discuss these three parts separately in the following sections.

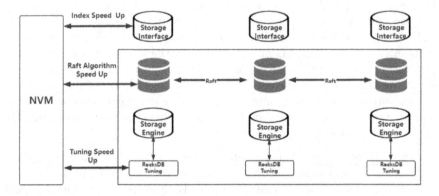

Fig. 1. Knowledge Graph Booster (KGB) structure

3.1   Boosting Neighbors Query

When processing graph data in graph database, queries often start from a vertex and continuously calculate the surrounding vertices and edges. In order to support the storage of hyper scale graph data and its related data processing, we designed and implemented a booster in persistent memory. This booster stores the neighbors ID of each vertex like index, with this booster, when querying the node's neighborhood first time, we store the one-hop results into index in persistent memory. Then for the following query, it will query in the index first, and it is much faster to find the one-hop result in index by using the key as node's id. If the id is not in index, use original process to query neighbors without index, and then store the one-hop result into index for the following queries.

The total process of querying neighbors is as Fig. 2. First, the query engine gets query from console and parses it. Then, if the query is type of getting neighbor's ID, the query will invoke get neighbors executor correspondingly. The executor will use Thrift RPC to send requests to processor in storage layer. Then the processor in storage engine builds and executes plan, and database store engine executes KV operations in executor.

In Nebula Graph, to query one node's neighborhood, all edges need to be traversed. If there are some data structures like adjacent list in memory, it can directly obtain all neighbor's ID, which is stored in the adjacent list. However, in database which did not have any approach to store the adjacent structure, all edges are searched when querying for a node's neighborhood.

This booster stores neighbors of each vertex in persistent memory. In KV format, the key contains vertex id, and also an index label to distinguish the index KV with vertex and edge's KV. Moreover, the visit time is also stored in key for record and compare the frequency of the index query. In this way,

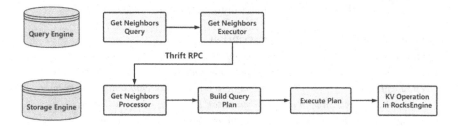

Fig. 2. Neighbors Query Process

if the space of index is not enough and some indexes needs to be deleted, the mechanism could delete indexes according to the visited time. For value part, it should contain the one-hop results as a vector. To expand this thought, two-hop or three-hop results can also be stored as the optimization plans.

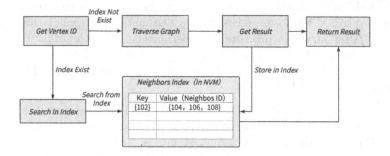

Fig. 3. Neighbors Query Process with Booster

As shown in Fig. 3, when traversing all edges' key in key-range, judge the source ID, and if it is equal to the query ID, the destination ID vertex is the neighborhood. Then the full answer will be obtained after traversing all keys in key range. It also should be noted that in the undirected graph, the edge's KV stores two times as source to destination and destination to source. So in both directed graph and undirected graph, this method can be handled by judging the source ID to get neighbors ID.

3.2 Performance Promotion with RocksDB

RocksDB is a famous LSM-tree based key-value store for SSDs which is derived form LevelDB [13]. Most modern distributed databases adopt RocksDB as their storage engine because of its lower write amplification and better write through-put performance. RocksDB offers a series of performance tuning options for users

Table 1. Performance tuning options in RocksDB.

Tuning Option	Description
filter_policy	*use bloom filters to reduce lookup cost*
block_cache	*use cache to promote read performance*
write_buffer_size	*determine the max write data amount in memory*
max_bytes_for_level_multiplier	*determine the LSM-tree's fan-out factor*
max_bytes_for_level_base	*determine the base level data amount*
...	...

to meet their specified application scenarios and current available resources. As shown in Table 1, we list a part of performance tuning options in RocksDB. We simply classify the performance tuning options into two main aspects, read and write.

Read Performance Tuning. Normally LSM-tree based key-value stores utilize bloom filters to check the existence of the target key to reduce the unnecessary lookup cost. Therefore, we could set a reasonable filter policy through the option of *filter_policy* to promote read performance. On the other hand, the cache also can speed up read performance by caching hot data blocks in memory. So users could set a larger *block_cache* size to cache more hot data blocks in memory instead of disk look-up.

Write Performance Tuning. As we all know, in LSM-tree based key-value stores, write data is first inserted into a Memtable that can access randomly. Then the Memtable is serialized into sorted string table (SSTable) files for persistence once the size reaches the threshold. So we could enlarge *write_buffer_size* option size to acquire higher throughput in short time. Meanwhile, *max_bytes_for_level_multiplier* and *max_bytes_for_level_base* can determine the LSM-tree's fundamental architecture, which could affect system's read and write performance respectively. There are still much more tuning options unlisted here for the length limitation. Based on the exact application scenario, the performance option tuning could achieve a magnificent promotion, which will be more precise to meet user's demand.

3.3   Raft Optimization in Distributed Knowledge Graph System

Raft is a distributed consensus algorithm. It has the same function as the paxos algorithm, that is, multiple nodes reach a consensus on a certain thing, even in the case of partial node failure, network delay, and network segmentation. Compared to paxos, the raft algorithm is simpler and easier to understand. Raft has been widely adopted by many practical distributed systems like Etcd, Nebula Graph, TiDB and CockroachDB since it was proposed in 2014.

In order to simplify, raft divides the algorithm into two parts: leader election and log replication. The raft algorithm will select one node as the leader among the nodes, and other nodes as the followers. The leader node interacts with the client and accepts the client's data requests, and then synchronizes the data to the follower. When the leader finds that more than half of the nodes save the data, it considers that the data has been committed, and applies it to its own state machine. Then respond to client requests. The data writing process of raft is shown in Fig. 4(a).

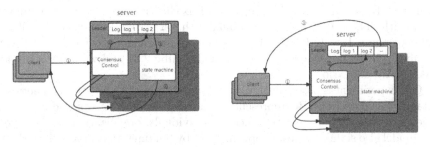

(a) The data writing process of raft   (b) Optimized write process

Fig. 4. Writing process of Raft

In the above data writing process, the apply of data is most likely to become the bottleneck of the system. Because the apply operation involves writing the state machine log to disk, sorting the data, and possibly writing the sorted data to disk, the apply operation is relatively slow. Therefore, when the client makes a large number of concurrent requests, there will be a large amount of request data in the system that has been committed but not applied. As a result, the request is suspended and cannot be returned. This greatly slows down the responsiveness of the system. So if we can parallelize the apply phase and the request return phase from Raft write processing, the time consumed will be greatly saved.

Therefore, we optimized the data writing process of raft. When raft replicates data to more than half of the nodes, we process the data apply and the response client in parallel. After the data is committed, we respond to the client immediately and execute the apply of the data concurrently and asynchronously. This is safe because the data is already stored on more than half of the nodes. The optimized raft writing process is shown in Fig. 4(b).

First of all, this saves the time for applying data and improves the efficiency of data writing for a single request. At the same time, a large number of requests that previously needed to wait for data application can be directly returned to the client without waiting, which greatly improves the response speed of the system.

4   Evaluation

We implement neighbors query index based on Nebula Graph v2.0.1 and Viper key-value store based NVM. We test get neighbors query of 1–4 hop and collect throughput, average latency and tail latency statistics.

4.1   Experimental Environment

We run experiments on a 64-core server equipped with Intel Xeo Gold 6242R CPUs clocked at 2.80 GHz with 44MB of L3 cache. The server is fully populated with 384 GB memory in total, occupying all six channels per socket for maximum bandwidth. The server runs Arch Linux with kernel 5.4.0-128-generic. All the code is compiled using GCC 9.4.0 with all the optimizations. Unless otherwise specified, all the experiments were run under glibc 2.31 and we use PMDK for PM/DRAM allocations.

Our experiment is based on (Linked Data Benchmark Council) social network dataset. The data in LDBC represents a snapshot of social network activity over time. The data includes entities such as individuals, organizations, and places. The model also simulates how people interact by building friendships with others, as well as sharing things like messages, including texts and images.

Our tests use "GO STEP" query in nGQL (query language used in Nebula), which traverses graph by filtered condition and returns the results of node's neighborhood. Our tests run on LDBC dataset with 1528 vertices and 14703 edges, and each query has virtual uses number of 5, and duration of 3 s.

4.2   Throughput Evaluation

We test the system throughput with booster and compare with the throughput with original Nebula system. Because booster in our system stores the previous results of neighbors, it can apparently speed up following queries. The promotion of performance increases largest in second repeat query, and gets smaller in next 3–7 queries.

In Fig. 5, our test also shows that the query with more hops gets more improvement in our system. Because in 1-hop without index, there is no index, and 2-hop or 3-hop, although there is no index at start, some index will be constructed when query, and used in following hops. Although the index is not constructed at first, it still helps a lot when constructing in the query process.

4.3   Query Latency Evaluation

As shown in Fig. 6, in our system, latency is decreased based on our booster. In 2-hop query, latency is decreased about 17%, and in 3-hop and 4-hop query, latency is decreased much more, about 78% and 89% respectively. With the hop number increasing from 2-hop to 4-hop, the response time delays about 7 times in original system, which is from 6000 µs to 43000 µs. However, in our system, with the index constructing, 3-hop and 4-hop latency can be decreased as same as 2-hop.

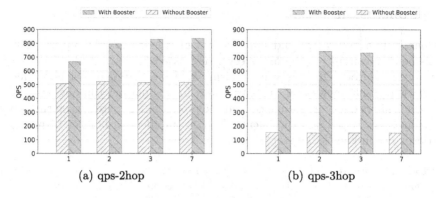

(a) qps-2hop   (b) qps-3hop

Fig. 5. Throughput Evaluation

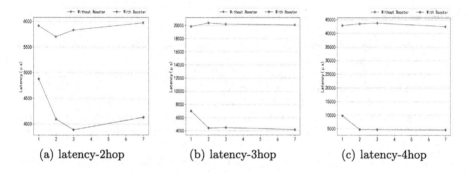

(a) latency-2hop   (b) latency-3hop   (c) latency-4hop

Fig. 6. Latency Evaluation

4.4   Tail Latency Evaluation

We also test p95 tail latency, and as Fig. 7 shows, our system also decreases tail latency greatly. In 2-hop query, tail latency is decreased by 40.8%, 3-hop and 4-hop are about 75%, and 81%. Therefore, our system can also decrease the latency of querying nodes with high connectivity.

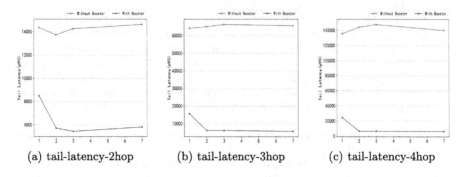

(a) tail-latency-2hop   (b) tail-latency-3hop   (c) tail-latency-4hop

Fig. 7. Tail Latency Evaluation

5   Conclusion

In our system, we design and implement a knowledge graph booster (KGB). This booster contains neighbors query index based NVM, raft algorithm optimization, and RocksDB tuning mechanism. We do evaluation on neighbors query index, and result shows that auxiliary index in this booster can obviously promote neighbors query performance, and the promotion increases when the query hops increase. Based on experiments, KGB is an effective booster to promote performance in knowledge graph system.

Acknowledgements. This work is supported by the National Key Research and Development Program of China (No. 2019YFE0198600), National Natural Science Foundation of China (No. 61972402, 61972275, and 61732014).

References

1. Sbadger: A fast key-value store written natively in go (2020). https://github.com/dgraph-io/badger
2. Balmau, O., Dinu, F., Zwaenepoel, W., Gupta, K., Chandhiramoorthi, R., Didona, D.: SILK: preventing latency spikes in log-structured merge key-value stores. In: 2019 USENIX Annual Technical Conference, pp. 753–766 (2019)
3. Buragohain, C., et al.: A1: a distributed in-memory graph database. In: Proceedings of the 2020 ACM SIGMOD International Conference on Management of Data, pp. 329–344 (2020)
4. Cao, Z., Dong, S., Vemuri, S., Du, D.H.: Characterizing, modeling, and benchmarking rocksdb key-value workloads at facebook. In: 18th {USENIX} Conference on File and Storage Technologies ({FAST} 2020), pp. 209–223 (2020)
5. Chai, Y., Chai, Y., Wang, X., Wei, H., Bao, N., Liang, Y.: LDC: a lower-level driven compaction method to optimize SSD-oriented key-value stores. In: 2019 IEEE 35th International Conference on Data Engineering (ICDE), pp. 722–733. IEEE (2019)
6. Chai, Y., Chai, Y., Wang, X., Wei, H., Wang, Y.: Adaptive lower-level driven compaction to optimize LSM-tree key-value stores. IEEE Trans. Knowl. Data Eng. **34**(6), 2595–2609 (2020)
7. Dageville, B., et al.: The snowflake elastic data warehouse. In: Proceedings of the 2016 International Conference on Management of Data, pp. 215–226 (2016)
8. Dayan, N., Athanassoulis, M., Idreos, S.: Monkey: optimal navigable key-value store. In: SIGMOD, pp. 79–94. ACM (2017)
9. Dong, S., Callaghan, M., Galanis, L., Borthakur, D., Savor, T., Strum, M.: Optimizing space amplification in rocksdb. In: CIDR (2017)
10. Gao, S., et al.: Formal verification of consensus in the taurus distributed database. In: Huisman, M., Păsăreanu, C., Zhan, N. (eds.) FM 2021. LNCS, vol. 13047, pp. 741–751. Springer, Cham (2021). https://doi.org/10.1007/978-3-030-90870-6_42
11. Gupta, P., Mhedhbi, A., Salihoglu, S.: Columnar storage and list-based processing for graph database management systems. Proc. VLDB Endow. **14**(11), 2491–2504 (2021)
12. Jain, M.: Dgraph: synchronously replicated, transactional and distributed graph database. birth (2005)
13. Leveldb - a fast and lightweight key/value database library by google (2017). http://code.google.com/p/leveldb

14. Li, S., Chen, W., Liu, B., Liu, P., Wang, X., Li, Y.-F.: OntoSP: ontology-based semantic-aware partitioning on RDF graphs. In: Zhang, W., Zou, L., Maamar, Z., Chen, L. (eds.) WISE 2021. LNCS, vol. 13080, pp. 258–273. Springer, Cham (2021). https://doi.org/10.1007/978-3-030-90888-1_21
15. Matsunobu, Y., Dong, S., Lee, H.: Myrocks: LSM-tree database storage engine serving Facebook's social graph. Proc. VLDB Endow. **13**(12), 3217–3230 (2020)
16. Miller, J.J.: Graph database applications and concepts with neo4j. In: Proceedings of the Southern Association for Information Systems Conference, Atlanta, GA, USA, vol. 2324 (2013)
17. Under the hood: Building and open-sourcing rocksdb (2017). http://goo.gl/9xulVB
18. Ruta, N.J.: CuttleTree: adaptive tuning for optimized log-structured merge trees. Ph.D. thesis, Harvard University (2017)
19. Sarkar, S., Papon, T.I., Staratzis, D., Athanassoulis, M.: Lethe: a tunable delete-aware LSM engine. In: Proceedings of the 2020 ACM SIGMOD International Conference on Management of Data, pp. 893–908 (2020)
20. Urbani, J., Jacobs, C.: Adaptive low-level storage of very large knowledge graphs. In: Proceedings of The Web Conference 2020, pp. 1761–1772 (2020)
21. Wang, Y., Chai, Y.: vRaft: accelerating the distributed consensus under virtualized environments. In: Jensen, C.S., Lim, E.-P., Yang, D.-N., Lee, W.-C., Tseng, V.S., Kalogeraki, V., Huang, J.-W., Shen, C.-Y. (eds.) DASFAA 2021. LNCS, vol. 12681, pp. 53–70. Springer, Cham (2021). https://doi.org/10.1007/978-3-030-73194-6_4
22. Wang, Y., Wang, Z., Chai, Y., Wang, X.: Rethink the linearizability constraints of raft for distributed key-value stores. In: 2021 IEEE 37th International Conference on Data Engineering (ICDE), pp. 1877–1882. IEEE (2021)
23. Wu, F., Yang, M.H., Zhang, B., Du, D.H.: AC-key: adaptive caching for LSM-based key-value stores. In: 2020 {USENIX} Annual Technical Conference ({USENIX}{ATC} 2020), pp. 603–615 (2020)

Towards a Unified Storage Scheme for Dual Data Models of Knowledge Graphs

Yuzhou Qin, Xin Wang$^{(\boxtimes)}$, and Wenqi Hao

College of Intelligence and Computing, Tianjin University, Tianjin, China
{yuzhou_qin,wangx,haowenqi}@tju.edu.cn

Abstract. As an important cornerstone of artificial intelligence, the knowledge graph is one of the indispensable foundations of the new generation of artificial intelligence from perception to cognition. RDF graphs and property graphs are the two main data models of KGs, and various data management methods have been developed for the two models. However, differences in data models will lead to differences in how the data is stored and manipulated, which will further hinder the widespread application of knowledge graphs. In this paper, we propose a novel unified storage scheme, UniS, which considers the characteristics of the two data model comprehensively. Unlike the existing approaches, the detailed conversion process for different storage forms of data that we devised will make it easier to manage multiple KGs in one database. Meanwhile, the experimental results show that UniS improves the storage and query efficiency by up to an order of magnitude than the state-of-the-art storage engines.

Keywords: Knowledge graphs · Data models · Unified storage scheme

1 Introduction

With the growing application of knowledge graphs in diverse domains, the scale of knowledge graphs (KGs) is dramatically increasing. As the two dominant data models for KGs, RDF (Resource Description Framework) [1] graph and property graph are utilized by most graph databases. For one thing, RDF, which has become a recommended standard for the representation of knowledge graphs by the World Wide Web Consortium (W3C), is widely adopted by databases represented by gStore [2] and Virtuoso [3]. For another, property graphs have been widely used as the data model by graph databases such as JanusGraph [4] and HugeGraph [5].

In recent years, there has been a consensus to unify the data model in the management of knowledge graphs, as different data models will lead to many differences in storage schemes and the databases built on them. Apart from the fact that different data models can cause problems for database users, existing storage schemes also have problems such as excessive storage overhead and null values. Therefore, we propose a unified storage scheme for RDF graphs and property graphs, which can accommodate both of them. The contributions of this paper can be summarized as follows:

(1) In order to store RDF graphs and property graphs, a unified storage scheme is proposed, i.e., UniS, which considers the characteristics of the two data model comprehensively. With UniS, the entities and edges are clustered and stored in separate tables according to their types, managing data in a unified way.

(2) To manage multiple KGs in one database, we have designed a detailed conversion process for different storage forms of data, which is convenient for further operation of data on this basis and meeting the storage and query load requirements of KGs.

(3) Extensive experiments are carried out to verify the effectiveness and efficiency of UniS. The experimental results show that UniS outperforms the state-of-the-art methods in terms of storage overhead and query overhead.

2   Related Work

In this section, we discuss the related works, including storage schemes and graph databases.

2.1   Storage Schemes

Most of the current storage solutions for RDF and property graphs are relationship-based. Triple table, adopted by 3store [6], stores data into tables with a 3-column structure, where each column corresponds to the subject, predicate, and object of a triple, respectively. Developed from triple table, horizontal table and property table are implemented in DLDB [7] and Jena [8], but they can lead to problems with excessive number of tables and null values. To reduce the join overhead of tables, the system represented by Hexastore [9] adopts sextuple indexing, but it will increase the amount of space required significantly. For the storage of property graph, storage schemes in the form of native graphs or key-value pairs are also utilized. Therefore, it is necessary to develop a unified and efficient storage scheme for RDF and property graphs.

2.2   Graph Database

gStore is an RDF graph database developed for storing triples of data, which utilizes the signature graph corresponding to RDF data and builds VS tree indexes to speed up SPARQL query processing. Moreover, supporting multiple data models including RDF data, Virtuoso is a hybrid database management system with built-in SPARQL and inference. On the other hand, designed for storing property graph, both JanusGraph and HugeGraph are also compatible with the query language Gremlin [10], but the Gremlin implementation cannot migrate directly from JanusGraph to HugeGraph due to the limitations of edge labels on HugeGraph.

3   Preliminaries

In this section, we introduce the definitions of relevant background knowledge.

Definition 1 (RDF Graph). *Let U and L be the disjoint infinite sets of URIs and literals. Then, a tuple t in the form of $\langle s, p, o \rangle \in U \times U \times (U \cup L)$ is called an RDF triple, where s is the subject, p the predicate, and o the object. An RDF dataset T, which is a finite set of triple t, can be converted to an RDF Graph $G = (V, E, \Sigma)$. The V, E, and Σ denote the set of vertices, edges, and edge labels in G, respectively. Formally, $V = \{s \mid (s, p, o) \in T\} \cup \{o \mid (s, p, o) \in T\}$, $E = \{(s, o) \mid (s, p, o) \in T\}$ and $\Sigma = \{p \mid (s, p, o) \in T\}$.*

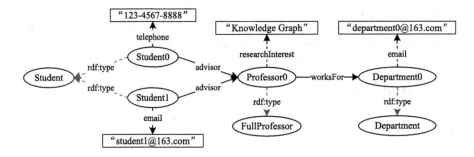

Fig. 1. An RDF graph example

Example 1. An example RDF graph G is shown in Fig. 1, which is composed of resources, associated properties and the relationships between resources. The ellipses and rectangles are used to denote resources and literals, respectively, while directed edges connecting vertices, which corresponds to the triples in the RDF dataset, represent relationships between vertices. In particular, the edge label `rdf:type` is employed to specify the type to which the resource belongs. For instance, the triple (`Professor0`, `rdf:type`, `FullProfessor`) indicates that the type of `Professor0` is `FullProfessor`.

Definition 2 (Property Graph). *Given a property graph $G = (V, E, \eta, src, tgt, \lambda, \gamma)$, where V, E represent the finite set of vertices and edges respectively, and $V \cap E = \emptyset$. The function $\eta : E \to (V \times V)$ denotes the mapping of edge to vertex pair, e.g., $\eta(e) = (v_1, v_2)$ means there is a directed edge e between vertex v_1 and vertex v_2. Moreover, the function $src : E \to V$ and $tgt : E \to V$ denote the mapping of edges to starting vertices and ending vertices, respectively, e.g., $src(e) = v_1$ denotes the starting vertex of edge e is v_1 and $tgt(e) = v_2$ denotes the ending vertex of edge e is v_2. Furthermore, The function $\lambda : (V \cup E) \to Lab$ represents the mapping of vertices or edges to labels, where Lab denotes the set of labels, e.g., let $v \in V$ (or $e \in E$) and $\lambda(v) = l$ (or $\lambda(e) = l$), then l is the label of vertex v (or edge e). In addition, the function $\gamma : (V \cup E) \times K \to Val$ represents the mapping of the associated property to a vertex or an edge, where K is the* *set properties and Val is the set of values, e.g., $v \in V$ (or $e \in E$), pro $\in K$ and $\gamma(v, pro) = val$ (or $\gamma(e, pro) = val$) denotes the value of the property pro on vertex v (or edge e) is val.*

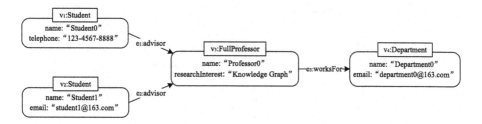

Fig. 2. A property graph example

Example 2. As shown in Fig. 2, every vertice and edge in the property graph has a unique id, and both vertices and edges have labels, e.g., the label `Student` on vertex v1 represents that the type of `Student0` is `Student`. Furthermore, Both vertices and edges have attributes, each of which consists of a key-value pair of an attribute name and an attribute value. For example, the attribute `researchInterest` on vertex v_3 is `Knowledge Graph`.

4   Unified Storage Scheme

In this section, we present a unified data model, named UniS, which is capable of storing both RDF and property graphs. We first introduce the design of the UniS storage model, followed by a description of the process for transforming RDF and property graphs into the UniS format.

4.1   Unified Storage Model

To store RDF graphs and property graphs in a unified manner, we propose a unified data model, which is combined with the characteristics of both the two data models.

UniS is based on the relation model, which is a tuple (R^v, R^e, N, μ), where R^v and R^e represent the set of **entities** and **edges** tables, respectively. For each entity table $r^v \in R^v$, r^v consists of two columns, where the first column records the globally unique identifier of each entity and the second column records the properties of entities. Meanwhile, for edge table $r^e \in R^e$, r^e contains three columns, storing the identifier of source entities, target entities, and properties of each edge, respectively. $N = \{n_1, n_2, ..., n_k\}$ is the set of table names, and the function $\mu : R^v \cup R^e \rightarrow N$ maps relation table r to its corresponding name.

As shown in Fig. 3, UniS is compatible with both property and RDF graphs. For the RDF graphs, we divide the entities into different entity tables based on

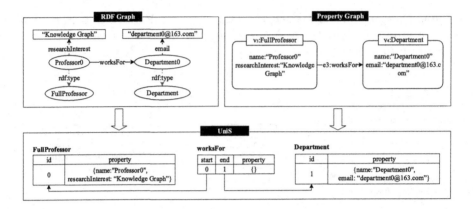

Fig. 3. The storage schema of UniS

the types specified by `rdf:type` and store the constant properties corresponding to the entities in the property column. The RDF graph in Fig. 3 contains two entities, i.e., `Professor0` and `Department0`, whose types are `FullProfessor` and `Department`, respectively, so we create the entity tables `FullProfessor` and `Department` in UniS, and then insert the entity identifiers and properties into them. For the edge `worksFor` between two entities, we create the edge table `worksFor` in UniS for it and insert the identifiers of source entity and target entity, and the properties of the edge into the edge table. For property graphs, similar approach can be adopted to transform them into the UniS. The details of the transformation process will be discussed in Sect. 4.2.

4.2 The Process of Transformation

To accommodate both RDF graphs and property graphs, the UniS are proposed to store both of them. However, due to the differences in the data model between RDF and property graph, various processes need to be designed to transform them into the unified storage schema.

For the storage of RDF graphs, the set of all triples T can be divided into 3 subsets: T_t, T_p, and T_e, which represent the set of triples related to the types of entities, the properties of entities, and the relationship between the entities, respectively. Formally, $T_t = \{t = (s, p, o) \mid t \in T \wedge p = \texttt{rdf:type}\}$, $T_p = \{t = (s, p, o) \mid t \in T \wedge o \in L\}$, and $T_e = \{t = (s, p, o) \mid t \in T \wedge p \neq \texttt{rdf:type} \wedge o \notin L\}$. In order to transform the triples to the unified storage model, further processing are necessary for the above sets of triples with the following conversion rules:

1. For any triple $t \in T_t$, the entity s should be put in the entity table whose name is o.
2. For any triple $t \in T_p$, the key-value pair (p, o) is inserted into the property column corresponding to entity s in the entity table.

3. For any triple $t \in T_e$, the edge will be put in the edge table named p as a record, where the *start* is s and the *end* is o.

The detailed processing flow is shown in Algorithm 1. The first step is to divide the RDF dataset T into three disjoint subsets T_t, T_p, and T_e (line 1–3). To locate the type and properties of each entity in an efficient way, we first divide the subjects in T_t by rdf:type (line 4–5) and the triples in T_p into groups by subject (line 7–8), then assign globally unique identifiers for all subjects in T_t (line 9–10). After that, we employ *props* to record the properties (line 13–15) for each entity and insert (id(s), *props*) into the corresponding entity table (line 16). For T_e, we employ a similar approach to handle it. First, the triples in T_p are grouped by the predicate (line 18–19). Then, for each triple $t = (s, p, o)$ in the group $\mathbb{E}(p)$, $(id(s), id(o), \emptyset)$ will be inserted into the edge table r^e (line 20–22), where $\mu(r^e) = p$.

Algorithm 1: Transform RDF Graph

Input: RDF dataset T

Output: $UniS = (R^v, R^e, N, \mu)$

1 $T_t \leftarrow \{t = (s, p, o) \mid t \in T \wedge p = \text{rdf:type}\}$;

2 $T_p \leftarrow \{t = (s, p, o) \mid t \in T \wedge o \in L\}$;

3 $T_e \leftarrow \{t = (s, p, o) \mid t \in T \wedge p \neq \text{rdf:type} \wedge o \notin L\}$;

4 **foreach** $t = (s, p, o) \in T_t$ **do**

5 $\quad \mid \quad \mathbb{S}(o) \leftarrow \mathbb{S}(o) \cup \{s\}$;$\qquad\qquad$ // group subjects by its type

6 $\quad \mid \quad N \leftarrow N \cup \{o\}$;

7 **foreach** $t = (s, p, o) \in T_p$ **do**

8 $\quad \mid \quad \mathbb{P}(s) \leftarrow \mathbb{P}(s) \cup \{t\}$;$\qquad\qquad$ // group triples by subject

9 **foreach** $s \in \{s \mid (s, p, o) \in T_t\}$ **do**

10 $\quad \mid \quad id(s) \leftarrow$ a globally unique identifier ;

11 **foreach** $n \in N$ **do**

12 $\quad \mid \quad$ **foreach** $s \in \mathbb{S}(n)$ **do**

13 $\quad \mid \quad\quad \mid \quad$ **foreach** $(s, p, o) \in \mathbb{P}(s)$ **do**

14 $\quad \mid \quad\quad \mid \quad\quad \mid \quad props(p) \leftarrow o$;$\qquad$ // insert (p,o) into properties

15 $\quad \mid \quad\quad \mid \quad props(\text{``uri``}) \leftarrow s$;

16 $\quad \mid \quad\quad \mid \quad r^v \leftarrow r^v \cup \{(id(s), props)\}$;$\qquad$ // insert data into entity table

17 $\quad \mid \quad R^v \leftarrow R^v \cup \{r^v\}$; $\mu(r^v) \leftarrow n$;

18 **foreach** $t = (s, p, o) \in T_e$ **do**

19 $\quad \mid \quad \mathbb{E}(p) \leftarrow \mathbb{E}(p) \cup \{t\}$;$\qquad\qquad$ // group triples in T_e by predicate

20 **foreach** $p \in \{p \mid (s, p, o) \in T_e\}$ **do**

21 $\quad \mid \quad$ **foreach** $t = (s, p, o) \in \mathbb{E}(p)$ **do**

22 $\quad \mid \quad\quad \mid \quad r^e \leftarrow r^e \cup \{(id(s), id(o), \emptyset)\}$;$\quad$ // insert edge data into edge table

23 $\quad \mid \quad R^e \leftarrow R^e \cup \{r^e\}$; $\mu(r^e) \leftarrow p$; $N \leftarrow N \cup \{p\}$;

24 **return** (R^v, R^e, N, μ) ;

The time complexity of Algorithm 1 is $O\left(|T| \cdot \log(|T|)\right)$, where $|T|$ is the numbers of triples in RDF dataset. The time complexity of the algorithm consists of two parts: (1) traverse triples in T_t, T_p, and T_e to generate mappings $\mathbb{S}$, $\mathbb{P}$, and $\mathbb{E}$, respectively, with complexity of $O\left(|T| \cdot \log(|T|)\right)$; (2) traverse triples in $\mathbb{S}$ and $\mathbb{E}$ to insert data into corresponding entity or edge tables, with complexity of $O(|T|)$. Hence, the overall time complexity of this algorithm is $O\left(|T| \cdot \log(|T|)\right)$.

For the storage of property graphs, it is relatively easy to transform them to the unified storage model as the property graph provides built-in support for properties on vertices and edges. Specifically, the data of vertices or edges with different labels in the property graph will be converted to records in the corresponding entity tables or edge tables, then their properties are stored in the properties columns. For instance, let $\lambda(v) = l$ (or $\lambda(e) = l$) and $\gamma(v, pro) = val$ (or $\gamma(e, pro) = val$), the entity v (or the edge e) should be inserted into the entity table (or the edge table) named l, and the (pro, val) will be put in the property column corresponding to entity v (or the edge e).

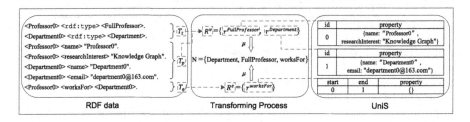

RDF data   Transforming Process   UniS

Fig. 4. The process of UniS

Example 3. Figure 4 shows the process of transforming RDF data to UniS. The set of triples is first classified according to the structure of them, then R^v can be obtained by further processing T_t and T_p, while R^e can be constructed according to T_e. Specifically, based on the type and attribute information of `Professor0`, the entity table $r^{FullProfessor}$ can be constructed. Moreover, the type `FullProfessor` is added to the set N and the mapping function μ between $r^{FullProfessor}$ and `FullProfessor` is built. In the same way, we can construct the entity table $r^{Department}$ and the edge table $r^{worksFor}$, and further obtain the final result of UniS.

5   Experiments

In this section, to verify the efficiency of the unified storage scheme, i.e., UniS, we compare it against the RDF databases gStore [2], Virtuoso [3] and the property graph database JanusGraph, HugeGraph on different datasets.

5.1   Experimental Settings

On the top of Nebula Graph, the proposed unified storage scheme is implemented and deployed on a 4-node cluster, which has a 16-core Intel(R) Xeon(R) Silver

4216 2.10 GHz CPU, 512 GB of RAM, and 1.92 TB SSD, running the 64-bit CentOS 7.7 operating system.

Data sets. Our experiments are conducted on the datasets of LUBM [11] and LDBC-SNB [12]. The LDBC Social Network Benchmark (SNB) models the social network graph, including people and their activities over time. The Lehigh University Benchmark (LUBM) is developed for evaluating the performance of Semantic Web repositories, which describes universities, departments, and their activities. In our experiment, we generate several different scales of data for both datasets.

Baselines. To verify the efficiency of storage, we compare the unified storage scheme with gStore, Virtuoso, JanusGraph, and HugeGraph in terms of both storage space overhead and loading time. Specifically, gStore 0.3.0 and Virtuoso 7.2.6 are RDF graph databases and JanusGraph 0.5.3 and HugeGraph 0.11.2 are property graph databases. Moreover, based on the unified storage scheme, experiments of query test was carried out to verify the query efficiency.

5.2 Experimental Results

Exp 1. Storage Efficiency. As shown in Fig. 5(a) and Fig. 5(b), UniS is significantly more efficient than gStoreD in terms of storage time and space for the LUBM dataset. UniS has a similar storage space overhead compared to Virtuoso, and the storage time of UniS is longer than Virtuoso when the data volume is small. However, as the data volume increases, UniS outperforms Virtuoso by up to an order of magnitude in loading data. For the LDBC dataset, as shown in Fig. 5(c) and Fig. 5(d), the storage efficiency of UniS is better than JanusGraph and HugeGraph.

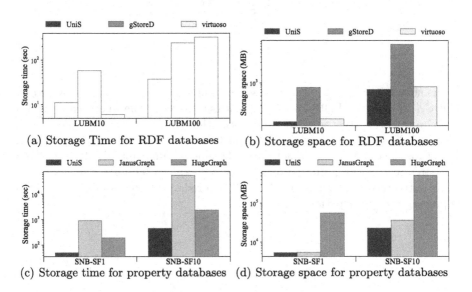

(a) Storage Time for RDF databases   (b) Storage space for RDF databases

(c) Storage time for property databases   (d) Storage space for property databases

Fig. 5. The experimental results of storage efficiency

There are two reasons for these results: (1) we devise an efficient transformation process where the time complexity is $O(|T| \cdot \log(|T|))$, which improves the efficiency of data loading significantly; (2)We employee advanced compression techniques, including dictionary encoding and common prefix extraction, to optimize the storage of large RDF datasets in UniS. By implementing these methods, we're able to compress the raw data and save on storage space, thus enabling UniS to store more data without requiring additional storage resources.

Exp 2. Query Efficiency. We executed the eight queries[1] over the LUMB10 dataset to verify the query efficiency of UniS on the RDF dataset. As can be seen in Fig. 6, the average query efficiency of UniS is 4.26 and 8.35 times higher than that of gStoreD and Virtuoso, respectively, for the following reasons: (1) UniS stores data of different types separately, which significantly accelerate the data filtering for queries of specified types. (2) UniS stores entities and their properties together, eliminating several join operations and improving the performance of queries about multiple properties of a single entity, such as Q6. (3) UniS compress the raw data, improving query efficiency by alleviating the burden of the disk.

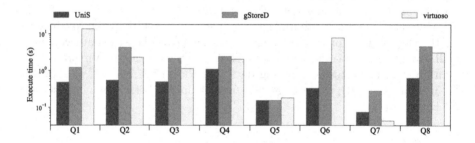

Fig. 6. Execute time on LUBM10

As shown in Fig. 7, we executed seven interactive short queries provided by LDBC [12] over the LDBC-SNB SF1 dataset to verify the query efficiency of UniS on the property graphs. It can be seen that the average query efficiency of UniS is 1.28 and 1.75 times that of JanusGraph and HugeGraph, respectively, except for the queries that do not finish. The most significant advantage of UniS compared to other property databases is that it exploits dictionary encoding and prefix extraction to compress the raw data to reduce the cost of disk read, thus improving the query performance.

---

[1] https://github.com/rainboat2/KGMA2022.git.

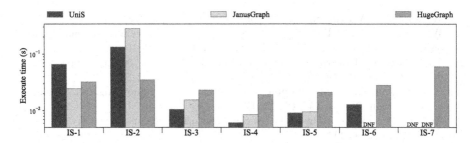

Fig. 7. Execute time on LDBC-SNB SF1 (DNF denotes dooes not finish within 30 min.)

6 Conclusion

In this paper, we propose UniS, a unified storage scheme for different data models on knowledge graphs. Considering the characteristics of RDF graphs and the property graphs comprehensively, the unified storage model is utilized to store the two data models in a unifed way. Furthermore, a detailed conversion process is devised, which makes it easier to manage multiple KGs in one database. Microbenchmarks over LUBM and LDBC are proposed to verify the efficiency and effectiveness of UniS. The experimental results show that UniS outperforms the state-of-the-art methods in terms of storage overhead and query overhead by up to an order of magnitude.

Acknowledgments. This work is supported by the National Key Research and Development Program of China (2019YFE0198600) and National Natural Science Foundation of China (61972275).

References

1. Consortium, W.W.W., et al.: Rdf 1.1 concepts and abstract syntax (2014)
2. Das, S., Agrawal, D., El Abbadi, A.: G-store: a scalable data store for transactional multi key access in the cloud. In: Proceedings of the 1st ACM Symposium on Cloud Computing, pp. 163–174 (2010)
3. Erling, O., Mikhailov, I.: Rdf support in the virtuoso dbms. In: Pellegrini, T., Auer, S., Tochtermann, K., Schaffert, S. (eds.) Networked Knowledge - Networked Media. Studies in Computational Intelligence, vol. 221. Springer, Berlin. pp. 7–24 (2009) https://doi.org/10.1007/978-3-642-02184-8_2
4. Authors, J.: Janusgraph–distributed graph database. http://janusgraph.org/ (2020)
5. Team, T.H.: The hugegraph manual. https://hugegraph.github.io/hugegraph-doc/ (2020)
6. Harris, S., Gibbins, N.: 3store: Efficient bulk rdf storage (2003)
7. Pan, Z., Heflin, J.: Dldb: extending relational databases to support semantic web queries. Lehigh univ bethlehem pa dept of computer science and electrical engineering, Technical Report (2004)

8. Wilkinson, K., Wilkinson, K.: Jena property table implementation (2006)
9. Weiss, C., Karras, P., Bernstein, A.: Hexastore: sextuple indexing for semantic web data management. Proc. VLDB Endowment **1**(1), 1008–1019 (2008)
10. TinkerPop, A.: Tinkerpop3 documentation v.3.3.3. http://tinkerpop.apache.org/docs/3.3.3/reference/ (2018)
11. Guo, Y., Pan, Z., Heflin, J.: Lubm: a benchmark for owl knowledge base systems. J. Web Seman. **3**(2–3), 158–182 (2005)
12. Angles, R., et al.: The ldbc social network benchmark. arXiv preprint arXiv:2001.02299 (2020)

Cross-platform Product Matching Based on Knowledge Graph

Wenlong Liu[1,2], Jiahua Pan[2], Xingyu Zhang[2], Xinxin Gong[1,2], Yang Ye[2],
Xujin Zhao[2], Xin Wang[3], Kent Wu[1], Hua Xiang[1], and Qingpeng Zhang[1,2(✉)]

[1] The Laboratory for AI -Powered Financial Technologies, Hong Kong SAR, China
`qingpeng.zhang@cityu.edu.hk`
[2] School of Data Science, City University of Hong Kong, Hong Kong SAR, China
[3] College of Intelligence and Computing, Tianjin University, Tianjin, China

Abstract. Product matching aims to identify similar or identical products sold on different platforms, which is crucial for retailers to adjust investment strategies. By building knowledge graphs (KGs), the product matching problem can be converted to the Entity Alignment (EA) task, which aims to discover the equivalent entities from diverse KGs. This paper introduces a two-stage pipeline consisted of rough filter and fine filter. Based on product names and categories, we roughly match products in rough filtering. For fine filtering, a new framework for Entity Alignment, **R**elation-aware, and **A**ttribute-aware Graph Attention Networks for **E**ntity **A**lignment (RAEA), is employed. Experiments on eBay-Amazon dataset indicated that the two-stage pipeline performs well on the problem of cross-platform product matching.

Keywords: Product matching · Knowledge graph · Entity alignment

1 Introduction

Identifying the same products originating from different e-commerce platforms, called product matching task [1], is one of the key challenges in e-commerce applications. E-commerce platforms build knowledge graphs with their own massive accumulated data. The product matching can be converted to entity alignment (EA) task among cross-language heterogeneous e-commerce knowledge graphs.

There are growing numbers of EA techniques based on GNNs recently since GNNs suit KGs' inherent graph structure. Commonly, graph structure [2], neighborhood information [3], relations and attributes are used to product matching [4]. More advanced approaches leverages multi-relation information between entities. These methods ignore the implicit interaction among relations and attributes since they do not integrate alignment signals from relations and attributes. Here, we employ a two-stage pipeline to match products from eBay and Amazon, where the RAEA model, a gnn-based EA model we proposed, is utilized as the fine filter. We find that the proposed model makes use of attributes, relations and the interaction among them, thus we achieve excellent product matching results based on product knowledge graphs.

© The Author(s), under exclusive license to Springer Nature Singapore Pte Ltd. 2023
S. Yang and S. Islam (Eds.): APWeb-WAIM 2022 Workshops, CCIS 1784, pp. 45–48, 2023.
https://doi.org/10.1007/978-981-99-1354-1_5

2   Data Processing

The Amazon product metadata used in our experiments contains descriptions, titles, categories, brands and prices for 957 thousand products [5]. The eBay product dataset includes name, supplier, price and categories for 46 thousand products.

From structured data, we extract product titles and categories. Other attributes such as product colors, size and target customer are extracted from unstructured text data based on regular expression rules. Since the data from eBay is in Chinese, following Sun et al. [4], we use google translate to translate Chinese data of eBay product to English. We build two KGs for eBay and Amazon respectively by constructing relation triples and attribute triples, which will be used in the fine filtering stage.

3   Methodology

To align Amazon products with eBay products, we take the eBay product as the query condition, search for the matched Amazon products, and sort the candidate results of the query. As the flowchart is shown in Fig. 1, all work is divided into three parts. Firstly, we extract product attributes from semi-structured data with Paddle UIE [7], which are combined with the structured data to build the KGs. The two-stage pipeline for product matching includes rough filtering and fine filtering, from where we get the product similarity matrix. Finally, the top-k most matched Amazon products for each eBay product are output after the post-processing. In this section, the two-stage pipeline is described in detail.

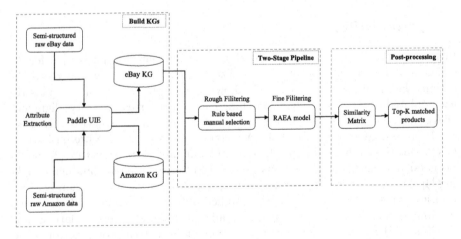

Fig. 1. Flow chart for product matching cross Amazon and eBay.

3.1   Rule-based Rough Filtering

To narrow down the number of amazon candidates matched to each eBay product, we use a rough filter based on regular expressions rules. According to the observation and statistics, the granularity of the stratification is not the same on both platforms, which is hard to align categories directly. Therefore, for each product, the hierarchical categories and title is concatenated to get a string containing basic information of the product. In this way, the matched Amazon candidate products contain roughly similar categorized features with the eBay product. The matching rules are designed manually based on expert knowledge.

3.2   Fine Filtering Based on Entity Alignment

For Amazon candidates of each eBay product, we try to select the most matched top-k products with the proposed EA model, RAEA. Designed by introducing relation-aware graph attention networks derived from RAGA [6] into the AttrGNN model [4], the RAEA captures interactions of entities and relations, and remains the diversity of relation types between two entities. As the framework of the RAEA model is shown in Fig 2, the graph is divided into four subgraphs based on structure and types of attribute value, representing different perspectives for alignment signals. Then, the four channels are trained respectively with the attributed value encoder and the relation-aware aggregator, where the attribute triples are aggregated to each product entity, and product embeddings are propagated in terms of relation triples. Finally, we ensemble the output from different channels to construct the final similarity matrix of products, from where the top-k most matched products are selected as final outputs.

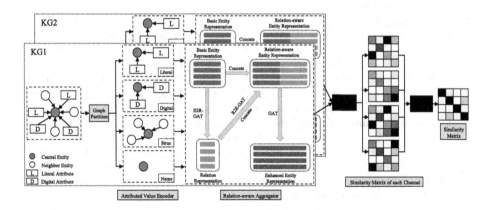

Fig. 2. The framework of RAEA model.

4   Experiments and Conclusion

We apply the two-stage pipeline to the dataset of eBay-Amazon products to observe the effectiveness of our method in real-world data. We mainly employ the NDCG as the evaluation matrix since the ranking measurement of candidate products reflects the performance of our method. If we match the top-10 most similar products, the NDCG is 0.58, the recall is 0.87 and the precision is 0.56, which means the candidates of matched Amazon products for each eBay product are kind of acceptable.

In this work, we employ a two-stage pipeline to match similar products that originate from two different e-commerce platforms, eBay and Amazon. In the rough filtering stage, the products are roughly filtered based on rules defined by categories and titles of products. In the fine filtering stage, we propose the RAEA model to find the top-k most matched Amazon products for each eBay product and achieve impressive results on the eBay-Amazon dataset.

Acknowledgments. The study in this paper was supported partly by the National Key Research and Development Program of China (No. 2019YFE0198600), and partly by InnoHK initiative, The Government of the HKSAR, and Laboratory for AI-Powered Financial Technologies.

References

1. Peeters, R., Bizer, C., Glavas, G.: Intermediate training of bert for product matching. small **745**(722), 2–112 (2020)
2. Wang, Z., Lv, Q., Lan, X., Zhang, Y.: Cross-lingual knowledge graph alignment via graph convolutional networks. In: EMNLP, pp. 349–357 (2018) https://doi.org/10.18653/v1/D18-1032
3. Yu, D., Yang, Y., Zhang, R., Wu, Y.: Knowledge embedding based graph convolutional network. In: Proceedings of the Web Conference 2021, pp. 1619–1628 (2021)
4. Liu, Z., Cao, Y., Pan, L., Li, J., Chua, T.-S.: Exploring and evaluating attributes, values, and structures for entity alignment. In: EMNLP, p. 10 (2020)
5. Ni, J., Li, J., McAuley, J.: Justifying recommendations using distantly- labeled reviews and fine-grained aspects. In: EMNLP-IJCNLP, pp. 188–197 (2019)
6. Zhu, R., Ma, M., Wang, P.: RAGA: relation-aware graph attention networks for global entity alignment. In: Karlapalem, K., Cheng, H., Ramakrishnan, N., Agrawal, R.K., Reddy, P.K., Srivastava, J., Chakraborty, T. (eds.) PAKDD 2021. LNCS (LNAI), vol. 12712, pp. 501–513. Springer, Cham (2021). https://doi.org/10.1007/978-3-030-75762-5_40
7. Paddle UIE Homepage. https://github.com/PaddlePaddle/PaddleNLP/tree/develop/model_zoo/uie. Accessed 1 Oct 2022

The Fourth International Workshop on Semi-structured Big Data Management and Applications

A GCN-Based Sequential Recommendation Model for Heterogeneous Information Networks

Dong Li, Yuchen Wang, Tingwei Chen[✉], Xijun Sun, Kejian Wang, and Gaokuo Wu

School of Information, Liaoning University, Shenyang 110036, China
dongli@lnu.edu.cn

Abstract. Sequential recommendation is an important task in the research of recommendation methods for heterogeneous information networks, which is a technique of predicting next user behavior according to completed behavior features. Traditional recommendation methods for heterogeneous information networks only consider a kind of user behavior feature, ignoring the heterogeneity of user behavior in heterogeneous information networks and the correlation relationship among user behavior. In this paper, we propose a sequential recommendation method based on multi-behavior features fusion (MBFF) for heterogeneous information networks, which makes full use of the heterogeneity and temporal features of user behavior to fuse multi-behavior features and effectively ensure the effectiveness of sequential recommendation. We propose an embedding fusion method of graph convolutional networks (GCN) to achieve the fusion of different behavior features. Moreover, A multi-head attention embedding fusion method based on dynamic heterogeneous networks is proposed to realize the fusion of user embedding and behavior embedding. Our method has experimented on real datasets, and the feasibility and effectiveness of our proposed method are demonstrated via experiments.

Keywords: Sequential Recommendation · Heterogeneous Information Networks · Multi-Behavior Features Fusion · Graph Convolutional Networks (GCN) · Attention Mechanism

1 Introduction

With the development of Internet, the online e-commerce platform has become an important platform for people to shop. Users and items in the online e-commerce platform are different nodes, these nodes and the association relationships between nodes together form a complex information network [1, 2]. In the massive data, although the recommendation system [3] can be used to recommend suitable items for users, how to combine deep learning and other new technologies to improve the accuracy of personalized recommendations [4] is a key problem. Sequential recommendation is a recommended method which uses the user's historical behavior sequence data to explicitly model to predict the user's next behavior and improve the recommendation effect [5].

In real-world information networks, most recommendation methods only consider a single behavior feature (eg. Purchase) between user and item, ignoring the heterogeneity

S. Yang and S. Islam (Eds.): APWeb-WAIM 2022 Workshops, CCIS 1784, pp. 51–62, 2023.
https://doi.org/10.1007/978-981-99-1354-1_6

and mutual influence of these behavior features, and cannot use relationships between various user behaviors. Heterogeneous information networks include nodes and complex associations between nodes, such as interactive networks and social networks composed of users and items. As shown in Fig. 1, a user has multiple behaviors such as "click", "add to shopping cart" and "purchase". User A and user C add a laptop to the shopping cart, respectively. User B and user C click and browse a tablet. User A buys a mobile phone, and user B buys a laptop. The multi-behavior features are that user A "buys" a mobile phone and "adds a laptop to the shopping cart.

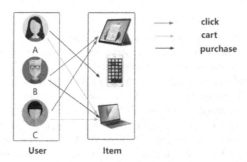

Fig. 1. An example of recommendation for heterogeneous information networks

In this paper, we present a sequential recommendation method based on multi-behavior features fusion (MBFF) for heterogeneous information networks. And we make the following contributions.

(1) Different from traditional recommendation methods, our MBFF is modeled in an end-to-end manner. Multi-type user-item interaction behaviors and the relationship between potential items make full use of the heterogeneity and temporal features of user behaviors to fuse multi-type user behavior features.
(2) According to the behavior feature embedding representations of different users, an embedding fusion method of graph convolutional network (GCN) is proposed to realize the feature extraction of homogeneous behaviors.
(3) For interaction features between users and various behaviors, we propose a multi-head attention embedding fusion method based on dynamic networks to realize the fusion of user embedding and behavior embedding.

2 Related Work

Many scholars have conducted research on recommendation methods, and recommendation algorithms are applied to many scenarios of social life, such as online e-commerce platform and social media.

The most widely used recommendation method is based on collaborative filtering [6]. In [7], Markov chains was used to model transitions in behavioral sequences. To enhance the ability to learn complex behaviors, the methods based on deep learning [8] had been proposed, including the methods based on recurrent neural networks (RNN)

[9] and the methods based on attention mechanisms. Some research work was also carried out from the perspective of network embedding. In [10], the deepwalk model was proposed. In [11], the node2vec model was designed to obtain information about the neighborhood of nodes. To further capture the heterogeneity of graphs, [12] proposed a meta-path random walk-based approach for a recommendation, which further improved the correlation between semantics. The input data was represented by a graph structure, and [13, 14] were used in the recommendation system. To decompose the user-item rating matrix into user and item embedding matrices [15] for the standard GCN, Wang et al. [16] further proposed a general solution for the implicit recommendation task. In [17], the graph convolutional neural network was improved, and the LightGCN model was proposed. The basic idea of these GCN-based recommendation models was to utilize higher-order neighbors in the user-item graph to refine the embeddings of target nodes by aggregating the embeddings of neighbors [18]. HAN [19] combined meta-paths to propose a hierarchical attention mechanism based heterogeneous graph convolutional network. [20] combined similarity-based methods with Markov chains for personalized sequence prediction on sparse and long-tail datasets.

3   MBFF Model

To make full use of the heterogeneity and timing of user behavior, we propose sequential recommendation MBFF model based on the fusion of multi-type behavior features for heterogeneous information networks.

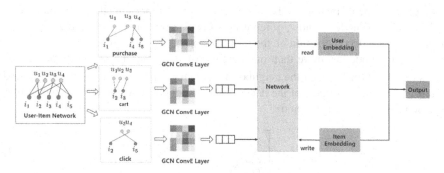

Fig. 2. Overview of MBFF model

3.1   Model Overview

The basic idea of the MBFF model is shown in Fig. 2. First, we combine the node information in the heterogeneous information networks, including user nodes and item nodes, to transform the loose item sequence into a tight user-item network and use it as the input of the model. Second, the MBFF model uses GCN and average pooling to extract features from nodes in heterogeneous information networks. Finally, the processed nodes serve as the base embedding for the fusion of heterogeneous memory networks.

3.2 Feature Extraction

A GCN embedding method is proposed for the embedding representation of different user behavior features to realize feature extraction of homogeneous user behaviors. We embed the nodes in the user-item network, and perform GCN convolution on them to discover their preferences for items. After the convolution layer is formed, the convolution layer is pooled to prevent overfitting.

Given a heterogeneous information network $G = (V, E)$, where V is the set of nodes and E is the set of edges between nodes. This paper studies the possible purchases of the remaining candidate items based on the historical behavior sequence of each user.

Let $U = \{u_1, u_2, \cdots, u_m\} \in V$ and $I = \{i_1, i_2, \cdots i_n\} \in V$ be the set of m users and the set of n items, respectively. Define N as the set of all neighbor nodes, W as the parameter matrix, and l as the interaction type. Users and items are nodes in the network, and edges represent user behavior. Since the number of node neighbors in an information network is not fixed, we build a GCN to process graph data.

The linear convolution method is used to calculate the nodes and extract the features of homogeneous user behavior, which can greatly reduce the complexity of calculation, helping solve the over-smoothing problem and improve the accuracy of recommendations.

When a node in each layer has only a single neighbor, its convolution representation is shown in (1).

$$v_l^{(k+1)} = \triangle f v_l^{(k)} w_l^k \tag{1}$$

l is a certain type of interaction between nodes, $v_l^{(k+1)}$ represents the convolution of k + 1 neighbor of node v, and $\triangle f$ represents Laplace operator. Laplace transform can perform feature decomposition on the user's historical sequence behavior.

When the node has more than one neighbor node, make the node self-connect. It needs to extract the order of the neighbor node under homogeneous behavior $v_{u,l}^{(k)}$, and multiply it by the reciprocal of the degree of the neighbor node, and add it to the order representation of the current node.

$$v_{u,l}^{(k+1)} = \left[\sum_{i \in N_u} \frac{1}{\sqrt{N_u}\sqrt{N_i}} v_{i,l}^{(k)} \right] v_{u,l}^{(k+1)} \tag{2}$$

As shown in (2), N_u indicates the degree of node u neighbor nodes. The degree of the node is used to express the user preference for the item.

$$v_{u,l} = \sum_{k=0}^{k} r_k v_{u,l}^{(k)} \tag{3}$$

As shown in (3), $v_{u,l}$ represents the aggregated embedding representation of nodes under the interaction relationship l. k represents the number of layers of convolutional convolutions of neighbors around the node, and r_k is a constant.

4   Interaction Sequences Fusion

For the interaction features between users and multi-type behaviors, a multi-head attention embedding fusion method based on dynamic heterogeneous networks is proposed to realize the fusion of user embedding and behavior embedding. The method of graph attention mechanism is used to assign weights to the embeddings of different behavior features, and the homogeneous behavior features obtained are fused to obtain the final representation of the node.

The item embedding q_i is given, and read and write operations are performed to predict the next user behavior sequence and recommend preferred item. We use a method similar to global feature F to extract features for multi-type behavior sequences, and use global features f_z to represent the global feature. Then read and write it.

The reading of multi-type behavior features is shown in (4) and (5).

$$\beta_{iz} = q_i^T f_z \tag{4}$$

$$m_u^k \leftarrow m_u^k + \beta_{iz} add_i \tag{5}$$

β_{iz} represents the parameters of global feature extraction for item embedding, and m_u^k represents the matrix module of the network. It calculates the embedding of the m-th user based on the parameters v_u^m.

The writing process of multi-type user behavior features is shown in (6) and (7). First, the matrix update for the current input record, because the user may have some noise records in the whole sequence. Second, it uses the tanh function to process the product embedding add_i. Third, use the parameters and update the matrix module.

$$add_i = \tanh(q_i) \tag{6}$$

$$m_u^k \leftarrow m_u^k + \beta_{iz} add_i \tag{7}$$

5   Recommendation Algorithm

5.1   Training and Prediction

We pass the inner product of the current user embedding and the item embedding by sigmoid function for calculating the prediction score of the recommendation system (defined as (8)).

$$y_{ui} = predict(v_u, q_i) = \sigma\left(v_u^T, q_i\right) \tag{8}$$

To prevent the model from overfitting, the normative parameter and the minimized loss function can be defined as (9).

$$Loss = -\frac{1}{|O|} \sum_{o \in O} \left(y_{ui} \log \widehat{y_{ui}}\right) + \lambda \|\theta\|_\mu^2 \tag{9}$$

O is the training set, and $|O|$ is the number of training instances. $y_{ui} = 1$ represents positive instances at that time and negative instances at that time. The positive and negative samples are randomly sampled by the idea of a random walk algorithm.

5.2 Algorithm Description

A sequential recommendation algorithm based on multi-type behavior features is proposed, as shown in Algorithm 1.

Algorithm 1. Recommendation Algorithm

Input: target nodes, embedding size K

Output: List of items with the highest rating

01. Nv = deepwalk(Nodes)

02.   while ($Loss < minLoss$)   // Iterative update

03.   for $i \leftarrow 1$ to N // Get links that do not exist in the matrix

04.   for $l \leftarrow 1$ to L // Homogeneous feature behavior embedding

05. $E[i][l] = Base(Nodes[i], l)$

06.   end for

07. $aggre(E[i])$

08.   end for   // Multi-class behavioral feature fusion embedding

09. $Loss = countLoss(v_u, v_i, i, u \in Nodes)$

10.   end while

11. for $i \leftarrow 1$ to N

12. $List[i] = topitem(Nodes[i])$

13. end for

The main steps of the algorithm are as follows.

Step 1. Initialize the data set, collect the generated target node set nodes and neighbor node set.

Step 2. Perform feature extraction on the target node as the input of the heterogeneous network.

Step 3. After implementing heterogeneous embedding, update the target node iteratively.

Step 4. Train the model, and calculate the recommendation score between user nodes and item nodes after prediction.

6   Experiments

We conduct experiments on two real datasets. By comparing our model with the traditional sequential recommendation model and non-sequential recommendation model, the effectiveness of our proposed model is verified.

6.1 Datasets and Experimental Settings

As shown in Table 1, we conduct experiments on two real datasets of online e-commerce platforms (the Jingdong dataset and the Taobao dataset).

Table 1. Datasets

Dataset	Type	Users	Items	Interactions
Jindong	cart	9890	136439	386452
	purchase	9890	21378	24576
	click	9890	18792	25675
Taobao	cart	9781	34908	10053
	purchase	9781	14589	36809
	click	9781	25091	28790

As shown in Table 2, the purchase behaviors of each user are divided into training sets and test sets, and we use single-type models and multi-type models to conduct experiments, respectively.

Table 2. The settings of single-typed models and multi-typed models

Models	Dataset	Positive Samples	Negative Samples
Single-typed	Training	Purchase	Never Purchase
	Test	Purchase	Never Purchase
Multi-typed	Training	Interact	Never Interact
	Test	Purchase	Never Purchase

We use HR, NDCG and MRR as indicators to evaluate the recommended performance.

(1) Hit rate (HR): Indicates the accuracy of predicting user behavior, and it is defined in (10).

$$HR = \frac{1}{N} \sum_{i=1}^{N} hits(i) \tag{10}$$

In (10), N represents the total number of users and $hits(i)$ represents whether the value is in the recommendation list, respectively.

(2) Normalized depreciation cumulative gain (NDCG): Evaluate the ranking results, and judge whether the items recommended by the user are placed in a more prominent position for the user.

(3) MRR: MRR is the average reciprocal rank, which is the average of the reciprocals of the first hit item rank.

6.2 Effectiveness Evaluation

We choose the following sequential recommendation model and non-sequential recommendation model to conduct comparative experiments with our proposed MBFF.

Non-sequential recommendation models are as follows.

(1) LightGCN. A graph neural network approach is used to improve the recommendation of high-order connectivity.
(2) DIN. An attention mechanism is used, and the user representation is obtained by aggregating historical interactions and attention weights.
(3) BPR. Bayesian-based recommendation model.

Sequence recommendation models are as follows.

(1) Caser. This method embeds a set of recent item sequences in time and latent space into an image feature and uses convolutional filters to learn its sequence patterns.
(2) GRU4REC. This method uses GRU to model the user session sequence and encodes the sequence of interest to the user as the final state.
(3) NextItNet. The network structure is formed by stacking multiple convolutional layers.

Table 3. Experimental results on the Taobao dataset

Model	MRR	NDCG@5	HR@5	NDCG@10	HR@10
LightGCN	0.0354	0.0232	0.0471	0.0377	0.0537
DIN	0.0205	0.0239	0.0476	0.0386	0.0494
BPR	0.0147	0.0154	0.0136	0.0203	0.0159
Caser	0.0547	0.0296	0.0563	0.0413	0.0667
GRU4REC	0.0612	**0.0315**	0.0291	0.0428	0.0342
NextItNet	0.0396	0.0203	0.0553	0.0463	0.0674
MBFF	**0.0676**	0.0308	**0.0657**	**0.0539**	**0.0712**

Table 4. Experimental results on the Jingdong dataset

Model	MRR	NDCG@5	HR@5	NDCG@10	HR@10
LightGCN	0.0466	0.0189	0.0574	0.0343	0.0815
DIN	0.0306	0.0194	0.0143	0.0252	0.0243
BPR	0.0159	0.0096	0.0132	0.0197	0.0218
Caser	0.0509	0.0251	0.0599	0. 0371	0.0671
GRU4REC	0.0471	0.0262	0.0267	0.0295	0.0315
NextItNet	0.0563	0.0175	0.0619	0.0329	0.0952
MBFF	**0.0606**	**0.0273**	**0.0632**	**0.0398**	**0.1031**

It can be seen from the above experimental results that most of the sequential recommendation models outperform the non-sequential models. The MBFF model proposed in this paper extends the original single interaction behavior sequence recommendation to a multi-type interaction behavior sequence recommendation. It can be inferred from Table 3 and Table 4 that multi-type interaction sequences can carry more useful information for users, especially users for the interaction information with the item, the performance is mostly improved compared to the baseline model. Therefore, the experimental results show that on the Taobao and the Jingdong datasets, our proposed MBFF model can more effectively recommend the user preferred items according to the user interaction sequence.

Table 5. Training time of models

Dataset	DIN	Caser	GRU4REC	NextItNet	MBFF
Taobao	21.37m	22.14m	24.91m	17.86m	17.27m
Jingdong	22.43m	40.61m	35.73m	26.54m	21.10m

Table 5 shows the training time of models on the two datasets. We can observe that our method is more than 20% more efficient than all baselines except the non-sequential model of DIN on the JD dataset. Therefore, we conclude that our proposed model can more effectively model the long-term historical sequence of users.

6.3 Ablation Analysis

To evaluate the influence of each component of our proposed MBFF model, we perform ablation analyses on the two datasets.

MBFF adopts a simplified graph convolution neural network in the first stage and graph attention mechanism in the second stage. We use only graph convolution or attention to conduct experiments in three different ways.

(1) MBFF-GG. That is, only the traditional GCN method is used in heterogeneous information networks.
(2) MBFF-TT. Only the graph attention mechanism is applied to the model.
(3) MBFF. Our MBFF model.

The experimental comparison of training effects is shown in Fig. 3. The horizontal axis and the vertical axis represent the sampling level and the corresponding HR value, respectively. It can be seen from the figure below that on the two datasets, the MBFF model is better than the other two models.

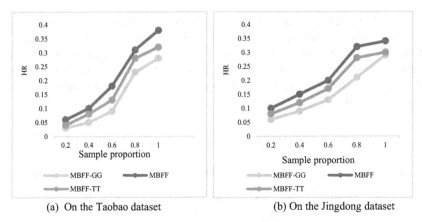

(a) On the Taobao dataset   (b) On the Jingdong dataset

Fig. 3. Comparison of training effects

6.4 Parameters Analysis

When recommending items by using the MBFF model, the influence of embedding dimension on the model is obvious. As shown in Fig. 4 (a) and (b), as the embedding dimension increases, the performance first grows rapidly and then gradually converges. Larger embedding dimensions increase the complexity of the model, while smaller dimensions lead to less accurate results. Therefore, we need to choose an appropriate embedding dimension and weigh the size of the embedding dimension, our model sets the embedding dimension to 60.

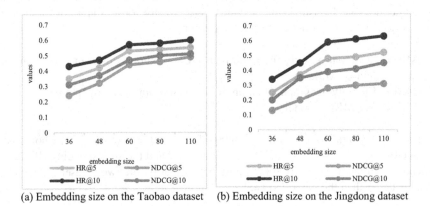

(a) Embedding size on the Taobao dataset   (b) Embedding size on the Jingdong dataset

Fig. 4. Embedding size analysis

7   Conclusions

In this paper, we propose a sequential recommendation method based on multi-behavior features fusion (MBFF) for heterogeneous information networks, which makes full use of the heterogeneity and temporal features of user behavior to fuse multi-behavior features and effectively ensure the effectiveness of sequential recommendation. Finally, experiments conducted on two real datasets demonstrate that our MBFF model can effectively solve sequential recommendation problem.

Acknowledgment. This work was supported by the Science and Technology Program Major Project of Liaoning Province of China under Grant No. 2022JH1/10400009, the Natural Science Foundation of Liaoning Province of China under Grant No. 2022-MS-171, 2020-BS-082, the Science Research Fund of Liaoning Province of China under Grant No. LJKZ0094, LJKQZ2021023.

References

1. Cui, P., Wang, X., Pei, J.: A survey on network embedding. IEEE Trans. Knowl. Data Eng. **31**(5), 833–852 (2019)
2. Li, D., Shen, D.R., Kou, Y., et al.: Information network fusion based on multi-task coordination. Front. Comp. Sci. **15**(4), 154608–154619 (2021)
3. Cui, Z., Xu, X., Xue, F., et al.: Personalized recommendation system based on collaborative filtering for IoT Scenarios. IEEE Trans. Serv. Comput. **13**(4), 685–695 (2020)
4. Wang, K., Zang, T., Xue, T., et al.: E-commerce personalized recommendation analysis by deeply-learned clustering. J. Vis. Commun. Image Represent. **71**, 102735 (2019)
5. Rendle, S., Freudenthaler, C., Schmidt-Thieme, L: Factorizing personalized Markov chains for next-basket recommendation. In: Proceedings of the 19th International Conference on World Wide Web, pp. 811–820 (2010)
6. Zhou, G., Mou, N., Fan, Y., et al.: Deep interest evolution network for click-through rate prediction. In: Proceedings of the 33rd AAAI Conference on Artificial Intelligence, pp. 5941–5948 (2019)
7. Dobrovolny, M., Selamat, A., Krejcar, O.: Session based recommendations using recurrent neural networks-long short-term memory. In: Proceedings of 33th Asian Conference on Intelligent Information and Database Systems, pp. 53–65 (2021)
8. Perozzi, B., Al-Rfou, R., Skiena, S.: Deepwalk: online learning of social representations. In: Proceedings the 20th ACM SIGKDD International Conference on Knowledge Discovery and Data Mining, pp. 701–710 (2014)
9. Grover, A., Leskovec, J: node2vec: scalable feature learning for networks. In: Proceedings of the 22nd ACM SIGKDD International Conference on Knowledge Discovery and Data Mining, pp. 855–864 (2016)
10. Dong, Y., Chawla, N.V., Swami, A.: metapath2vec: scalable representation learning for heterogeneous networks. In: Proceedings of the 23rd ACM SIGKDD International Conference on Knowledge Discovery and Data Mining, pp. 135–144 (2017)
11. Zhou, J., Cui, G.Q., Hu, S., et al.: Graph neural networks: a review of methods and applications. AI Open **1**, 57–81 (2020)
12. Busbridge, D., Sherburn, D., Cavallo, P., Hammerla, N.Y.: Relational graph attention networks. arXiv preprint arXiv:1904.05811 (2019)

13. Berg, R.V.D., Kipf, T.N., Welling, M.: Graph convolutional matrix completion. arXiv preprint arXiv:1706.02263 (2018)
14. Wang, X., He, X.N., Wang, M., et al.: Neural graph collaborative filtering. In: Proceedings of the 42nd International ACM SIGIR Conference on Research and Development in Information Retrieval, pp. 165–174 (2019)
15. He, X., Deng, K., Wang, X., et al.: LightGCN: simplifying and powering graph convolution network for recommendation. In: Proceedings of the 43rd International ACM SIGIR Conference on Research and Development in Information Retrieval, pp. 639–648 (2020)
16. Wu, S., Sun, F., Zhang, W., et al.: Graph neural networks in recommender systems: a survey. arXiv preprint arXiv:2011.02260 (2020)
17. Wang, X., Ji, H.Y., Shi, C.,et al.: Heterogeneous graph attention network. In: Proceedings of the World Wide Web Conference, pp. 2022–2032 (2019)
18. He, R., Julian, J.M.: Fusing similarity models with Markov chains for sparse sequential recommendation. In: Proceedings of IEEE the 16th International Conference on Data Mining, pp. 191–200 (2016)
19. Tang, J., Wang, K.: Personalized top-n sequential recommendation via convolutional sequence embedding. In: Proceedings of the 11th ACM International Conference on Web Search and Data Mining, pp. 565–573 (2018)
20. Yuan, F., Karatzoglou, A., He, X.N., et al.: A simple convolutional generative network for next item recommendation. In: Proceedings of the 12th ACM International Conference on Web Search and Data Mining, pp. 582–590 (2019)

SP-LAN: A Stock Prediction Model Based on LSTM-Attention Network

Jingyou Sun[1], Dong Li[1(✉)], Xing Wang[2], Yue Kou[2], Peixuan Li[1], and Yang Xie[1]

[1] School of Information, Liaoning University, Shenyang 110036, China
`dongli@lnu.edu.cn`
[2] School of Computer Science and Engineering, Northeastern University,
Shenyang 110004, China

Abstract. With the prosperity of the stock market and the rapid development of artificial intelligence, stock prediction technology has been widely concerned in recent years. Currently, most existing works only focus on stock prediction models via machine learning, ignoring some problems, such as local minima, gradient vanishing and interaction of information at different times. The stock data has characteristics like high noise and nonlinearity, which increase the difficulty of modeling. To solve these problems, we propose the Stock Prediction Model based on LSTM-Attention Network (called SP-LAN). Unlike traditional methods, SP-LAN can train the model with excellent generalization ability. SP-LAN also can make the model better learn interaction of information at different times, thus improving the prediction effectiveness of our model. In addition, we propose a novel feature selection strategy based on embedded importance to reduce the dimensions of data, which makes model train easier. Finally, we conduct experimental studies based on the daily trading dataset of SSE 50. The experimental results demonstrate the effectiveness and efficiency of our model.

Keywords: Stock Prediction · Feature Selection · LSTM · Attention Mechanism

1 Introduction

The stock market is an essential economic information center which produces a large amount of stock trading every day. If the data of stock trading is analyzed systematically, it will create a great deal of helpful information that dramatically benefits investors. With the development of the market economy, the stock market is becoming an important part of the financial industry. Many researchers generally recognize that a sound and effective stock market can predict the operation of the real economy, thus the stock prediction has fundamental practical significance. Because of the importance of the stock market to the economy, research on stock prediction was widely concerned in 1990s. In order to predict the trend of stock price, researchers apply various statistical and econometric methods on the research of the stock market. However, these methods have some drawbacks. With the development of artificial intelligence, researchers begin study stock prediction via machine learning and deep learning, and they tried to improve the accuracy of stock

S. Yang and S. Islam (Eds.): APWeb-WAIM 2022 Workshops, CCIS 1784, pp. 63–74, 2023.
https://doi.org/10.1007/978-981-99-1354-1_7

prediction by optimizing and combining different methods. The stock market is also affected by the external environment, the economic policies and the business status of the enterprise. The stock data has characteristics like high noise, nonlinearity and multi-dimensionality, which brings difficulties to the stock prediction. Traditional stock prediction methods have the following characteristics.

(1) Traditional statistical methods cannot overcome the uncertainty of the model caused by the high noise and nonlinear characteristics of stock data.
(2) Although the stock prediction model based on neural networks improves the accuracy of prediction, these methods ignore some problems, such as local minima, gradient vanishing and interaction of information at different times.
(3) Stock prediction is not only influenced by basic factors, such as opening price, highest price, lowest price, closing price and trading volume, but also affected by economic market environment, public opinion, government economic policies, corporate finance and investors' investment tendency.

To solve these problems, we propose the Stock Prediction Model based on LSTM-Attention Network (called SP-LAN). On the one hand, we train the model with good generalization capability. On the other hand, the model can learn the interaction of information at different times. More specifically, we make the following contributions.

(1) We review the related research works of stock prediction, and analyze the shortcomings of existing works.
(2) Different from traditional stock prediction models, our SP-LAN model introduces attention mechanism into LSTM to learn the mutual influence and interaction of information at different times, thus improving the prediction effectiveness of SP-LAN model.
(3) We propose a stock feature selection strategy based on embedded importance. This method is used to evaluate the importance of features comprehensively. On the one hand, we extract features that are more important to model calculation and remove features with less influence, thus shortening the time of model calculation. On the other hand, we reduce the impact of meaningless features on model prediction and improve the prediction effectiveness of the model effectively.

2   Related Work

Various approaches for stock prediction have been studied over the years. Traditional approaches mainly include interdisciplinary methods, BP methods, RNN methods and LSTM methods.

With the deepening of interdisciplinary, some techniques applied in other fields were constantly applied on stock prediction, such as wavelet analysis [1] and grey prediction method [2]. At the same time, researchers had begun to expand the methods of stock prediction to the field of artificial intelligence [3] to improve the accuracy of stock prediction via optimization and the combination of various methods [4].

In the process of continuous exploration, [5] proposed to improve the BP neural networks by using the PSO algorithm to enhance the reliability of the neural networks. In [6], the research proposed an improved bacterial chemotaxis optimization (IBCO), which was integrated into the BP neural networks to develop an efficient prediction model for predicting various stock indices. However, BP neural networks not only converges too slowly and easily falls into local minima, but also cannot consider the time series as feedforward neural networks.

Different from the BP neural networks, RNN have feedback links to learn the time pattern inside the neural networks, and it can build models and analyze the time series data effectively. In addition, RNN can also memorize historical input information and system state information. In [7], RNN were applied to stock prediction. RNN used the back-propagation algorithm when training the model, thus causing huge amounts of parameters. Because of the excessive number of layers, the training speed slowed down, and problems like gradient vanishing and gradient explosion occurred.

LSTM was proposed to improve the traditional RNN model, thus solving problems like gradient vanishing effectively. Compared with the BP neural networks, RNN and other neural networks, LSTM can not only predict nonlinear data but also learn time series information and solve the problem of gradient explosion effectively. Therefore, more and more researchers applied LSTM to stock prediction [8–11]. Sun [12] used LSTM to predict the price trend of the U.S. stock index and compared the prediction results with BP and RNN, which proved that LSTM could skip the local optimum and reach the global optimum.

With further research, researchers not only tried different neural network models but also put forward optimization and improvement schemes [13] aiming at improving the shortcomings of the models to improve the prediction effect. At the same time, influenced by many factors, such as the market economy environment, financial news [14, 15], public opinion [16], and investor sentiment [17, 18], the stock data presents the characteristics of high noise, nonlinearity and multi-dimensionality. Those characteristics caused a noticeable impact on the prediction effect. In order to reduce the impact caused by meaningless data, researchers began to process the input data via many scientific methods, such as text mining [19] and data dimensionality reduction [20].

3   Model Overview

In this paper, we propose the Stock Prediction Model based on LSTM-Attention Network (called SP-LAN), which introduces the self-attention mechanism into LSTM. Firstly, we introduce the background of the problem, summarizes the stock prediction process, and then introduce the construction, training, and optimization of our SP-LAN model in detail.

3.1 Background

With the development of big data, deep learning, and their application in various fields, the stock prediction method have begun to transition from the traditional statistical model to the neural network model. Traditional statistical methods, such as ARIMA and GARCH are all used to predict the stock by building appropriate mathematical models to fit the historical time trend curve. Therefore, traditional statistical methods cannot overcome the high noise of stock data and cannot learn the nonlinearity of stock data. In the neural network methods, the input variables of the BP neural network are single, ignoring the multidimensional nature of stock data in time series. RNN can learn the nonlinearity of stock data better, but it has the problems like gradient vanishing and gradient explosion. LSTM realizes the memory of time via the design of the gate and solves the problem of gradient vanishing of RNN. However, LSTM only filters the information of the current moment into the next moment every time through forget gate, and it ignores the connection and interaction between the information of each moment. Therefore, this paper puts forward a prediction model based on LSTM, which introduces the self-attention mechanism [21], so that this method can learn the relationship and influence between the information of different moments to improve the effect of stock prediction.

3.2 SP-LAN Model

Basic Idea. The information of the LSTM model at time t selectively transmits the information to time $t + 1$ through the forget gate. Although the memory function on time is realized, the mutual influence and interaction between information at different times are ignored in this process. Therefore, the self-attention mechanism is introduced into the last hidden layer to learn the interaction between information at different times to improve the effect of stock prediction. Figure 1 illustrates the framework of our SP-LAN model.

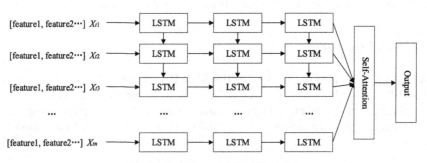

Fig. 1. Overview of SP-LAN model

x_t represents the input vector at time t, which is composed of data feature information at time t, and the cells of the model are LSTM cells. The data of the last hidden layer of the model is the output of the self-attention mechanism, and the final prediction result is the output through calculation.

Figure 2 is an internal structure diagram of the cell of LSTM, in which the trend of data can be seen. Where xt represents the input at time t, and h_{t-1} represents the output value at the previous time.

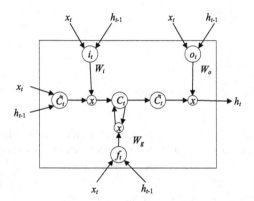

Fig. 2. Cell structure of LSTM

The important equations (as in (1)–(6)) of LSTM are given as follows.

$$\text{Forget gate: } f_t = \sigma\left(w_f * \left[h_{t-1}, x_t\right] + b_f\right) \tag{1}$$

$$\text{Input gate: } i_t = \sigma\left(w_i * \left[h_{t-1}, x_t\right] + b_i\right) \tag{2}$$

$$\text{Output gate: } o_t = \sigma\left(w_0 * \left[h_{t-1}, x_t\right] + b_o\right) \tag{3}$$

$$\text{Candidate cell state: } \widetilde{C}_t = \tanh(w_c * [h_{t-1}, x_t] + b_c \tag{4}$$

$$\text{Current cell state: } C_t = f_t * C_{t-1} + i_t * \widetilde{C}_t \tag{5}$$

$$\text{Output: } h_t = o_t * \tanh(C_t) \tag{6}$$

The self-attention mechanism is a special case of attention mechanism in that the value of the key is equal to the value. The calculation process is shown in Fig. 3.

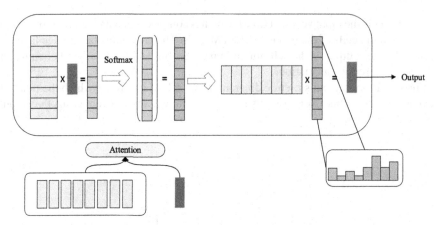

Fig. 3. Calculation process of self-attention mechanism

The values of h_{i1}, h_{i2}, ..., h_{it} are obtained via the calculation of LSTM cell in Fig. 2. These values are the output of i-th layer at time t, and then they are used as the input value of the self-attention mechanism. First, the similarity $F(H_i, h_{it}) = H_i^T h_{it}$ is calculated by point multiplication through h_{it} and H_i. H_{it} is the i-th layer output value at time t, and H_i is the output vector of the i-th layer. Second, the weight value a_t at time t is obtained via the softmax function. The specific formula is defined as (7).

$$a_t = Softmax(F(H_i, h_{it})) = \frac{exp(F(H_i, h_{it}))}{\Sigma_j exp(F(H_i, h_{ij}))} \tag{7}$$

The final attention output result is defined as (8).

$$Attention(H_i) = \Sigma_j a_t h_{ij} \tag{8}$$

The training of LSTM apply the back-propagation algorithm. There are mainly three steps in the training process.

Step1. Calculating the output value of each cell forward. For LSTM, it is the value of five vectors (f_t, i_t, C_t, o_t and h_t).

Step2. Calculating the value δ of the error term for each cell reversed. Like the recurrent neural network, the back-propagation of LSTM error term also includes two directions. The one is the back-propagation along time, that is, the error term of each time is calculated from the current time t. And the other one is to propagate the error term to the up level.

Step3. According to the corresponding error term, calculate the gradient of each weight.

Workflow. The workflow of our SP-LAN model includes data acquisition, data pre-processing, feature selection, model construction, model training and model prediction. The specific flow chart is shown in Fig. 4.

First, the stock data is acquired from the daily trading data. Second, the data is preprocessed, and the feature is selected through a novel feature selection strategy. In this process, Lagrangian interpolation is used to process the missing values. Abnormal data is detected by box chart, and then the abnormal value is processed by the average correction. Then, the SP-LAN model is constructed and trained. Finally, the training process is finished when the model converges, and the model is saved for data prediction.

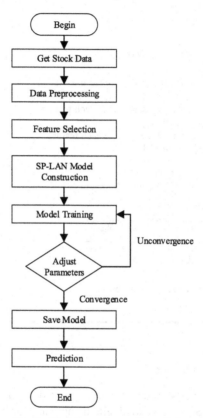

Fig. 4. Workflow of overall prediction

3.3   Feature Selection Strategy

Based on the multidimensional characteristics of stock data, we proposed a novel feature selection strategy based on embedded importance. This strategy is used to reduce the dimension of stock data. The feature selection algorithm is shown in Algorithm 1. It is mainly divided into three steps. Step 1 (lines 1–9) calculate the scores of DT (Decision Tree), RF (Random Forest), AdaBoost (Adaptive Boosting) and GBDT (Gradient Boosting Decision Tree) for each feature, respectively. Step 2 (lines 10–15) calculate the weight parameters α, β, γ and δ, then weight them with score vector to obtain the embedded importance. Step 3 (lines 16–19) compares the embedded importance with the threshold θ to decide whether retain the feature or not.

Algorithm 1. Feature Selection Algorithm

Input: training set $T = \{(x_1, y_1), (x_2, y_2), ..., (x_N, y_N)\}, x_i \in X, y_i \in Y$
      loss function $L(y, f(x))$

Output: the remaining features

1.   Calculation:
2. imp_{1i} // DT scores of each feature
3. $s_1 = [imp_{11}, imp_{12}, ... imp_{1n}]$
4. imp_{2i} // RF scores of each feature
5. $s_2 = [imp_{21}, imp_{22}, ... imp_{2n}]$
6. imp_{3i} // AdaBoost scores of each feature
7. $s_3 = [imp_{31}, imp_{32}, ... imp_{3n}]$
8. imp_{4i} // GBDT scores of each feature
9. $s_4 = [imp_{41}, imp_{42}, ... imp_{4n}]$
10. Calculation:
11. $\alpha = 1 - \dfrac{MSE_{DT}}{MSE_{DT} + MSE_{RF} + MSE_{AdaBoost} + MSE_{GBDT}}$
12. $\beta = 1 - \dfrac{MSE_{RF}}{MSE_{DT} + MSE_{RF} + MSE_{AdaBoost} + MSE_{GBDT}}$
13. $\gamma = 1 - \dfrac{MSE_{AdaBoost}}{MSE_{DT} + MSE_{RF} + MSE_{AdaBoost} + MSE_{GBDT}}$
14. $\delta = 1 - \dfrac{MSE_{GBDT}}{MSE_{DT} + MSE_{RF} + MSE_{AdaBoost} + MSE_{GBDT}}$
15. $Emb_{imp} = \alpha s_1 + \beta s_2 + \gamma s_3 + \delta s_4$
16. If $Emb_{imp}[i] \geq \theta$
17.   choose m_i feature as one of the features
18. Else
19.   delete m_i feature

3.4 Algorithm of SP-LAN Model

The algorithm of our SP-LAN model is mainly divided into three steps. Step 1 (lines 1–7) is to calculate the LSTM cell and obtain the output value of the hidden layer. Step 2 (lines 8–10) is to calculate the similarity between the output values of different cells in the LSTM hidden layer, then obtain a probability value between 0 and 1 via softmax function. Step 3 (line 11) is to weigh and sum the probability and the corresponding value to get the final prediction result via global information. The algorithm is shown in Algorithm 2.

Algorithm 2. Algorithm of SP-LAN Model

Input: feature matrix at time t

Output: prediction value at time $t+1$

1. Calculation:
2. $f_t = \sigma\big(w_f * [h_{t-1}, x_t] + b_f\big);$ // Forget gate
3. $i_t = \sigma(w_i * [h_{t-1}, x_t] + b_i);$ // Input gate
4. $\tilde{C}_t = \tanh(w_c * [h_{t-1}, x_t] + b_c;$ // Candidate cell state
5. $C_t = f_t * C_{t-1} + i_t * \tilde{C}_t;$ // Current cell state
6. $o_t = \sigma(w_0 * [h_{t-1}, x_t] + b_o);$ // Output gate
7. $h_t = o_t * \tanh(C_t);$ // Output
8. Calculation:
9. $F(H_i, h_{it}) = H_i^T h_{it};$ // The similarity between H_i and h_{it}
10. $a_t = \text{Softmax}\big(F(H_i, h_{it})\big) = \dfrac{exp\big(F(H_i, h_{it})\big)}{\Sigma_j exp\big(F(H_i, h_{ij})\big)};$ // The weight value of time t
11. $\text{Attention}(H_i) = \Sigma_j a_t h_{ij}.$ // Attention value (final output)

3.5 Optimization Strategy

Deep learning algorithm involves optimization in many cases. Different optimizers have different effects on different network models. Learning rate adjustment is a process which consumes a lot of computing resources. If the parameters are adjusted adaptively, the cost will be significantly reduced. Therefore, we use the Adam optimizer. Adam optimizer uses a method to calculate the adaptive learning rate for each parameter. It not only stores the exponential decay average of the past gradient square v_t, but also keeps the exponential decay average of the past gradient m_t. The formulas are defined as (9)–(13).

$$m_t = \beta_1 m_{t+1} + (1 - \beta_1)g_t \tag{9}$$

$$v_t = \beta_2 v_{t-1} + (1 - \beta_2)g_t^2 \tag{10}$$

$$\hat{m}_t = \frac{m_t}{1 - \beta_1^t} \tag{11}$$

$$\hat{v}_t = \frac{v_t}{1 - \beta_2^t} \tag{12}$$

$$\text{Update formula: } \theta_{t+1} = \theta_t - \frac{\eta}{\sqrt{\hat{v}_t} + \varepsilon}\hat{m}_t \tag{13}$$

The recommended default settings for the Adam optimizer are $\alpha = 0.001$, $\beta_1 = 0.9$, $\beta_2 = 0.999$ and $\varepsilon = 10^{-8}$.

4 Experiments

4.1 Datasets and Experimental Settings

In this paper, the daily trading data of the Bank of China and China Railway Construction in the SSE (Shanghai Stock Exchange) 50 Index are used as the dataset. The data from

January 2008 to December 2018 is used as the training set, and the data from January to July 2019 is used as the test set. MSE, RMSE, and MAE are used as prediction indicators in experiments. MSE means the average of the sum of squares of prediction errors. RMSE means the square root of MSE. MAE means the average of the sum of prediction errors.

As the datasets are different, the optimal parameters of different models will get different values. In the experiments, the parameters of the SP-LAN model are set as follows. The time window, the batch size, the epoch, the number of hidden cells and the number of hidden layers are 10, 50, 500, 22 and 3, respectively.

4.2 Effectiveness Evaluation

To evaluate our SP-LAN model, we choose DT, GBDT and AdaBoost are baseline models. The comparative experimental results on the Bank of China dataset are shown in Table 1.

Table 1. Experimental results on the Bank of China dataset

Model	MSE	RMSE	MAE
DT	0.00361	0.06012	0.04565
GBDT	0.00391	0.06255	0.04615
AdaBoost	0.00544	0.07378	0.05898
SP-LAN	**0.00272**	**0.05217**	**0.04043**

Compared with DT, GBDT and AdaBoost model, our model achieves the lowest MSE, RMSE and MAE of 0.00272, 0.05217 and 0.04043, respectively.

The comparative experimental results on the China Railway Construction dataset are shown in Table 2.

Table 2. Experimental results on the China Railway Construction dataset

Model	MSE	RMSE	MAE
DT	0.28437	0.53326	0.41858
GBDT	0.28219	0.53122	0.42000
AdaBoost	0.10313	0.32114	0.25165
SP-LAN	**0.06521**	**0.25536**	**0.18568**

Compared with DT, GBDT and AdaBoost model, our model achieves the lowest MSE, RMSE and MAE of 0.06521, 0.25536 and 0.18568, respectively.

5 Conclusions

In this paper, we propose the Stock Prediction Model based on LSTM-Attention Network (called SP-LAN). Different from the traditional stock prediction models, our model introduces the attention mechanism into LSTM to learn the interaction of information at different times. Moreover, we propose a novel feature selection strategy based on embedded importance to keep or delete features to reduce the dimension of the dataset. Finally, the experimental results demonstrate the effectiveness and efficiency of our stock prediction model.

Acknowledgement. This work was supported by the Science and Technology Program Major Project of Liaoning Province of China under Grant No. 2022JH1/10400009, the Natural Science Foundation of Liaoning Province of China under Grant No. 2022-MS-171, 2020-BS-082, the Science Research Fund of Liaoning Province of China under Grant No. LJKZ0094, LJKQZ2021023.

References

1. Li, T.: Research on stock fluctuation prediction method based on wavelet analysis and BP neural network. Tianjin University (2018)
2. Li, X.Q.: Research and application of grey prediction in stock price. Times Finance **10**, 158–160 (2017)
3. Tsiliyannis, C.A.: Markov chain modeling and prediction of product returns in remanufacturing based on stock mean-age. Eur. J. Oper. Res. **271**(2), 474–489 (2018)
4. Kristjanpolleri, R.W., Michell, V.K.: A stock market risk forecasting model through integration of switching regime, ANFIS and GARCH techniques. Appl. Soft Comput. **67**, 106–116 (2018)
5. Zhang, W.X.: BP neural network stock prediction model based on PSO optimization. Harbin Institute of Technology (2010)
6. Zhang, Y.D., Wu, L.N.: Stock market prediction of S&P 500 via combination of improved BCO approach and BP neural network. Expert Syst. Appl. **36**(5), 8849–8854 (2009)
7. Gao, T.W., Chai, Y.T.: Improving stock closing price prediction using recurrent neural network and technical indicators. Neural Comput. **30**(10), 2833–2854 (2018)
8. Naik, N., Mohan, B.R.: Study of stock return predictions using recurrent neural networks with LSTM. In: Proceedings of International Conference on Engineering Applications of Neural Networks, pp. 453–459 (2019)
9. Kim, H.Y., Won, C.H.: Forecasting the volatility of stock price index: a hybrid model integrating LSTM with multiple GARCH-type models. Expert Syst. Appl. **103**, 25–37 (2018)
10. Zhan, X.K., Li, Y.H., Li, R.X., et al.: Stock price prediction using time convolution long short-term memory network. In: Proceedings of International Conference on Knowledge Science, Engineering and Management, pp. 461–468 (2018)
11. Huang, B., Ding, Q., Sun, G.Z., et al.: Stock prediction based on Bayesian-LSTM. In: Proceedings of the 2018 10th International Conference on Machine Learning and Computing, pp. 128–133 (2018)

12. Sun, R.Q.: Research on price trend prediction model of U.S. stock index based on LSTM neural network. Capital University of Economics and Business (2016)
13. Tao, Z., Hou, M.Z., Liu, C.H.: Prediction stock index with multi-objective optimization model based on optimized neural network architecture avoiding overfitting. Comput. Sci. Inf. Syst. **15**(1), 211–236 (2018)
14. Tan, J.H., Wang, J., Rinprasertmeechai, D., et al.: A tensor-based eLSTM model to predict stock price using financial news. In: Proceedings of the 52nd Hawaii International Conference on System Sciences, pp. 1–10 (2019)
15. Matsubara, T., Akita, R., Uehara, K.: Stock price prediction by deep neural generative model of news articles. IEICE Trans. Inf. Syst. **101-D**(4), 901–908 (2018)
16. Zhang, G.W., Xu, L.Y., Xue, Y.L.: Model and forecast stock market behavior integrating investor sentiment analysis and transaction data. Clust. Comput. **20**(1), 789–803 (2017). https://doi.org/10.1007/s10586-017-0803-x
17. Liu, J., Lu, Z.C., Du, W.: Combining enterprise knowledge graph and news sentiment analysis for stock price prediction. In: Proceedings of the 52nd Hawaii International Conference on System Sciences, pp. 1–9 (2019)
18. Vanstone, B.J., Gepp, A., Harris, G.: The effect of sentiment on stock price prediction. In: Proceedings of International Conference on Industrial, Engineering and Other Applications of Applied Intelligent Systems, pp. 551–559 (2018)
19. Feuerriegel, S., Gordon, J.: Long-term stock index forecasting based on text mining of regulatory disclosures. Decis. Support Syst. **112**, 88–97 (2018)
20. Kondo, M., Bezemer, C.-P., Kamei, Y., Hassan, A.E., Mizuno, O.: The impact of feature reduction techniques on defect prediction models. Empir. Softw. Eng. **24**(4), 1925–1963 (2019). https://doi.org/10.1007/s10664-018-9679-5
21. Vaswani, A., Shazeer, N., Parmar, N., et al.: Attention is all you need. arXiv preprint arXiv: 1706.03762v5 (2017)

Research on Composite Index Construction Method Based on Master-Slave Blockchain Structure

Guiyue Zhang, Yu Sui, Jiacheng Zhang, Chen Liu, Wenlong Hu, and Tingwei Chen[✉]

School of Information, Liaoning University, Shenyang 110036, China
twchen@lnu.edu.cn

Abstract. Blockchain is a new information processing technology that uses efficient cryptographic principles for the trustworthy storage of big data. With the exponential growth of the scale of data on the chain, the problems of low query efficiency and long traceability time of the existing blockchain system become more and more serious. To solve the above problems, this paper proposes a composite index construction method based on a master-slave blockchain structure. Firstly, a weight matrix is introduced to slice the whole master-slave blockchain based on the master chain structure; secondly, a master index construction method based on jump consistency hash is proposed for the master blockchain within each slice; finally, based on an improved Bloom filter, a slave composite index is constructed for each master block corresponding to the slave blockchain. Experimental results show that, compared with the existing methods, the proposed method can reduce the index construction time by 4.63% on average, improve the query efficiency by 6.71%, and reduce the memory overhead by 24.4%.

Keywords: Blockchain · Indexing · Slice

1 Introduction

Blockchain stores and verifies data through a blockchain data structure, and ensures the security of data transmission and access with cryptography [1]. It has the characteristics of high credibility, traceability, and decentralization [2], and can solve the problem of trust of the third party in data storage [3]. With the development of blockchain technology and the accumulation of data in various industries, the traditional single-chain blockchain system has been unable to meet the increasingly complex application scenarios. Master-Slave Blockchain (MSBC) structures (such as Spark Chain, etc.) have begun to attract the attention of experts and scholars in the field, and are gradually widely used in education, medical care, security [4], and other fields [5, 6].

Master-slave blockchain usually includes a master chain and slave chain, which are composed of a master block and slave block respectively, and each master block has only one slave chain. Master-slave blockchain structure can cope with the application of complex classification scenarios. For example, in the financial field, the master-slave

S. Yang and S. Islam (Eds.): APWeb-WAIM 2022 Workshops, CCIS 1784, pp. 75–85, 2023.
https://doi.org/10.1007/978-981-99-1354-1_8

blockchain is used to build a company blockchain system that generates financial activities. The master chain stores the information of financial institutions, and the corresponding slave chains store their transaction events, financial activities, and other data. Through the consensus mechanism of blockchain [7], the data cannot be tampered with.

With the continuous increase of data scale, the problems of low query efficiency and long tracing time of blockchain systems [8, 9] become more and more serious. However, the existing blockchain indexes are only suitable for a single chain structure, and the query efficiency and dynamic maintainability are poor. Therefore, how to establish an efficient and dynamically maintainable index structure based on a master-slave blockchain has become a hot and difficult point in the field of research.

To solve the above problems, this paper proposes a composite index construction method (MSCI) based on master-slave blockchain structure. The main contributions of this paper are as follows:

(1) A weight matrix is constructed by integrating node load, node credit, and network quality, and a WMS master-slave blockchain fragmentation algorithm is proposed to realize the dynamic fragmentation of the master-slave blockchain structure based on the characteristics of the main chain.
(2) On this basis, aiming at the problem that the existing blockchain index is not suitable for the master-slave chain structure, an index construction method based on jump consistency hash is proposed on the master chain. The key value of the master chain node is mapped with the index slot to realize the efficient query of the master chain information.
(3) On the slave chain, combining the data characteristics of blockchain, an index construction method based on an improved Bloom filter is proposed.
(4) Compared with the existing methods on different constraints and data sets, the effectiveness of the proposed method is verified.

2   Related Work

At present, many scholars have conducted in-depth research on the index construction of blockchain and achieved certain research results.

Vikram Nathan et al. [10] proposed a multi-dimensional memory read optimized index (Flood) method, which automatically adapts to specific data sets and workloads by jointly optimizing the index structure and data storage layout. However, this method needs to evaluate the query cost of the current layout regularly, and completely rebuild the index for different workloads. Huang et al. [11] proposed the B-tree-based indexing (EBTree) method, which supports real-time top-k, range, and equivalent search of Ethereum blockchain data. However, the index nodes of this method are stored separately in Level DB, and the query efficiency is greatly affected by the node size. Andreas Kipf et al. [12] proposed a single-channel learning index RS (RadixSpline) method, which can be constructed in a single transfer of sorted data, and it is friendly to most data sets, but this method will reduce the performance with the increase of data sets. Xing et al. [13] proposed an index method of account transaction chain based on subchain (SCATC), which divided the transaction chain into sub-chains, and converts the query mode traversing the transaction chain into sub-chain query through pointers to reduce the computational overhead, but this method only improved the query efficiency of the longer account transaction chain and only optimized the query in plaintext. Noualhamdi et al. [14] put forward an extensible distributed index (ChainLink) method for large time-series data, designed a two-layer distributed index structure, and realized query operation with partition-level data reorganization. However, this method requires data reorganization locally, which can't guarantee data security. Gao Yuanning et al. [15] put forward a scalable learning index model (Dabble) based on the middle layer, which uses the K-Means clustering algorithm, divides the data set into K regions according to the data distribution, and uses a neural network to learn and train, respectively, and predicts the data position through the neural network model. However, this method requires retraining the model when updating the data set, and the timeliness of the model is poor, and the value of K has a great influence on the accuracy of the model.

To sum up, this paper makes an in-depth study on the index construction of the master-slave blockchain. Aiming at the shortcomings of existing index methods, and considering the space-time complexity and data security of index construction, a composite index construction method based on master-slave blockchain structure is proposed.

3   Blockchain Slice Method Based on Master-Slave Structure

To achieve efficient index construction and query processing of the master-slave blockchain structure, firstly, the whole master-slave blockchain is sliced based on the characteristics of the master chain, and each segment is given a weight. Based on this, the segmentation weight matrix of the whole master-slave blockchain structure is constructed, and the number of nodes in the slice is determined based on the weight matrix, which provides support for the indexing of the master chain and slave blockchain.

3.1   Construction of Weight Matrix

Let the number of nodes in the master-slave blockchain be x, the master-slave blockchain is divided into $y(y \ll x)$ slices, the i-th slice is $fi(i = 0, 1, 2, \cdots, y-1)$, and the weight of each slice is ω_i. The slice weight is determined by the weights of three dimensions: node load, node credit, and network quality, in which node credit and network quality are positively correlated with slice weight. Before calculating the slice weight, the units of the above three dimensions are not uniform, so normalization is needed.

Among them, the normalization formula of node load is:

$$d_{ij} = \left\lfloor \frac{1}{\log(d'_{ij} + 1) + 1} \times \frac{x}{y} \right\rfloor \tag{1}$$

The normalization formula of node credit and network quality is:

$$d_{ij} = \left\lfloor \frac{d'_{ij} + 1 - min}{max + 1 - min} \times \frac{x}{y} \right\rfloor \tag{2}$$

The weight of the *i-th* slice is ω_i, and the calculation formula is shown in formula (3).

$$\omega_i = \sum_{j-1}^{3} \omega_{ij} d_{ij} \tag{3}$$

d'_{ij} in formulas (1) and (2) represents the original values of node credit and network quality, d_{ij} is the normalized value, and *max* and *min* are the maximum and minimum values of this component, respectively. The ω_{ij} in formula (3) represents the weight of each component.

Let each element of the two-dimensional weight matrix M be composed of slice weights. After the weight of each slice is obtained, a two-dimensional weight matrix $M[p][q]$ is constructed by using each slice's weight ωi (where $p \leq \sqrt{y}$, and p and q are integers). For any slice f, there is $M[f/p][f\%q] = \omega_f$, and the empty position in the matrix is set to 0.

3.2 Determining the Number of Nodes in a Slice Based on Weight Matrix

Determine the number of nodes in each slice based on the weight matrix in Sect. 3.1. Firstly, the slice weights in the matrix are linearly normalized. Secondly, all normalized slice weights are dispersed in proportion, and the interval of discrete proportional weights is set as $[1, Q]$. The final slice proportional weight Q-1 is obtained, that is, the slice will correspond to Q-1 nodes.

Number the nodes in the order of $\{0, 1, 2, ..., \sum_{i}^{x-1} \omega_i - 1\}$, and the node number corresponding to the *kth* slice is $[\sum_{i=0}^{k} \omega_i, \sum_{i=0}^{k+1} \omega_i - 1]$. After obtaining the node number, the slice number can be obtained by looking up the table.

Among them, the linear normalization formula is:

$$\omega'_i = \frac{\omega_i - \omega_{min}}{\omega_{max} - \omega_{min}} \tag{4}$$

ω_{min} in formula (4) is the minimum value of all slice weights, ω_{max} is the maximum value, and ω'_i is the normalized slice weight.

Based on the weight matrix in Sect. 3.1, a weighting matrix-based slice algorithm (WMS) is proposed, which is shown in Algorithm 1 and realizes the slice of the master-slave blockchain structure.

Algorithm 1. WMS Algorithm

```
WMS(key,r,M[p][q]){
    int i ,j ;
    for i = 0 to p-1
      for j = 0 to q-1
      num_row = sum(M[i][j])
      row = jumpSearch(key,r,num_row);
      /*Call the jump search function*/
      row = row*q;
      for i = 0 to q-1
      num_col = sum(M[i][j])
      colmun = jumpSearch(key,r,col);
      /*Call the jump search function*/
      return(row,colmun);
}
```

4 Methods for Constructing Composite Index

Because the data scale and information types stored in the master chain and slave chain of the master-slave blockchain structure are different, after slicing by the WMS algorithm, a composite index construction method based on jump consistency hash and improved Bloom filter is proposed to meet the query requirements of the master chain and slave chain. The schematic diagram of composite index construction is shown in Fig. 1.

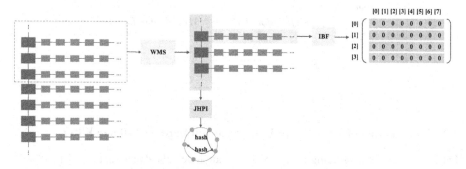

Fig. 1. Composite index construction schematic

4.1 Construction of Master Chain Index Based on Jump Consistency Hash

Based on the characteristics of data stored in the master chain, this paper introduces the jump consistent hash algorithm and proposes a jump consistent hashing-based master chain index construction method (JHPI) to realize the fast construction of the master chain index. Firstly, the number of index slots is determined according to the number of nodes in each master chain slice; secondly, the key value of each node is determined according to the hash value of the master chain stored data; finally, the key value of each node and the number of index slots are input, and the master chain index is output.

When the number of nodes in a slice changes, the nodes jump in the index, and some nodes are remapped. Let the hash mapping function that produces the jump change be $ch(key, num\_buckets)$, key be the key value of nodes, and $num\_buckets$ be the number of slots, when the number of slots changes from n to $n + 1$, the slot where the node with $n/(n + 1)$ is located remains the same, that is, $ch(key, num\_buckets) = n\text{-}1$, and $1/(n + 1)$ keys need to be remapped, that is, $ch(key, num\_buckets) = n$.

The b is the result of the last jump of the node, and j is the result of the next jump. For any $i(i \in [b + 1, j - 1])$, the probability that the number of nodes does not jump is shown in formula (5).

$$P(i) = \frac{b+1}{b+2} \times \frac{b+2}{b+3} \times \cdots \times \frac{i-1}{i} = \frac{b+1}{i} \tag{5}$$

The construction method of the master chain index based on jump consistent hash is shown in Algorithm 2.

Algorithm 2. JHPI Algorithm

```
int JumpHashIndex(int key, int num_buckets,int[] s) {
    int b = 1, j = 0;
    while (j < num_buckets) {
        b = j;
        key = key * 2862933555777941757ULL + 1;
        /* Generate random numbers with key and r as
seeds*/
        j = (b + 1) * (double(1LL << 31) / double((key >>
33) + 1));
    }
    return b;
}
```

4.2 Construction of Slave Chain Index based on Improved Bloom Filter

In the process of constructing the dependent chain index, the data structure of the Bloom filter is reconstructed, and the column-based selection function is proposed to realize the dependent chain index construction method (IBF) based on the improved Bloom filter.

Firstly, the data structure is a two-dimensional array $A[p][q]$, where $p = 2^n$ (n is a positive integer), and the data length in q is lq, assuming that $lq = 32/64$ bits. The value is determined by the cache line length of general registers in the CPU to reduce memory access and improve query performance. Let the K hash functions of the improved Bloom filter be $Hash(key)$, where the key is the key value of the node, and let the length of the elements that can be stored in an improved Bloom filter be len, and the calculation result of len is shown in formula (6).

$$len = p \times q = 2^n \times lq \tag{6}$$

After the data structure of the index is constructed, the index of the slave chain corresponding to each main block is constructed, and the specific steps are as follows:

Step1: Use the function of selecting columns, first map the element to the corresponding column, and the element will be at the position of the corresponding column;

Step2, obtaining the locus through K hash functions;

Step3: Set the corresponding site to 1, and the rest sites to 0.

Among them, the column selection function of Step1 is based on the SHA256 hash function, and the column selection function is obtained by modular operation. As shown in formula (7):

$$qv = v\%2^n \tag{7}$$

In Step2, the K hash functions are composed of K bitwise AND operations, which can be expressed as:

$$pv^i = v\& \sum_{j}^{i \times \frac{k'}{k} - 1} 2^{i-1} (0 < i \leq k) \tag{8}$$

The v in formula (7) is the element in the Bloom filter, and qv is the column number obtained after column selection. After obtaining the column in which the element is located, it is determined that the element will be limited to the corresponding column during construction and query. The pv^i in formula (8) is the row number obtained by K times of operation of formula (8) in the column after the column number is obtained, that is, the corresponding position, and k' is the array length in the general Bloom filter. The slave chain index construction algorithm based on the improved Bloom filter is shown in Algorithm 3.

Algorithm 3. IBF_Construction

$IBF\_Construction(v,k,) k' \{$
    $int\ p = 0, q = 0;$
    $for(v)\{$
      $q = v\ \%\ 2^n;$
      $p = v\ \&\ \sum_{j}^{i \times \frac{k'}{k} - 1} 2^{i-1};$
      $output\ p, q;$
    $\}$
$\}$

---

5 Experiments

The experimental environment of this paper is 16 servers with 1T storage space, 8G RAM, and a 4-core CPU. The servers communicate with each other through a high-speed network, and each server is equipped with an ubuntu 18.04 operating system. Two different data sets are used for experimental verification. The first data set is the first 3,000,000 blocks in the public Ethereum network, and there are 15,362,853 transactions in the data set. The second data set is the Lognormal artificial data set. Lognormal data set samples 5 million pieces of non-duplicate data according to lognormal distribution (mean value is 0, variance is 2). In this section, we will verify the high efficiency and low memory advantages of MSCI from three aspects: index building time, query time, and memory consumption.

5.1 Comparison of Index Construction Time

To verify the efficiency of the MSCI method proposed in this paper in terms of index construction time, the EBTree method with improved blockchain structure and Dabble model method with the neural network will be compared respectively. In the EBTree method, the capacity of internal nodes is set to 128, and the capacity of leaf nodes is set to 16. The value of k in Dabble model is 100, and the number of nodes in a slice of the MSCI method is set to 100. In this section, the experiment will be divided into three specific situations as shown in Table 1.

Table 1. Index build time comparison.

	Dataset One		Dataset Two	
	MSCI	EBTree	MSCI	Dabble
Case 1	The slave block data is empty, and the number of master blocks is 500k, 1000k, 1500k, 2000k, 2500k, and 3000k respectively		The slave block data is empty, and the master block stores 2, 4, 6, 8, and 100,000 pieces of data	
Case 2	The master block data is empty, and the number of slave blocks is 500, 1000, 1500, and 2000		The master block data is empty, and the slave blocks store 500, 1000, 1500, and 2000 pieces of data	
Case 3	The data of master and slave blocks are not empty, the number of master blocks is 500k, 1000k, 1500k, 2000k, 2500k, and 3000k respectively, and the number of slave blocks is 1000		The data of master and slave blocks are not empty. The master block stores 2, 4, 6, 8, and 100,000 pieces of data, and the slave block stores 1,000 pieces of data	

As shown in Table 1, the experimental results of index construction time comparison are shown in Fig. 2 and Fig. 3.

As can be seen from Fig. 2 and Fig. 3, with the increase in data volume, compared with the existing methods, the index construction time of MSCI is optimized by about 4.79% and 4.47% respectively.

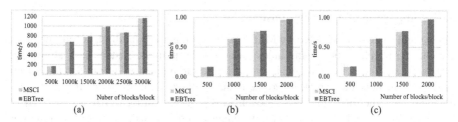

Fig. 2. Index build time comparison 1

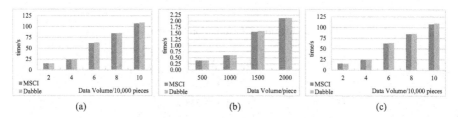

Fig. 3. Index build time comparison 2

5.2   Comparison of Query Time

In this experiment, to test the query performance of MSCI, we will compare the query response time of the EBTree method when the number of main blocks is 500k, 1000k, 1500k, 2000k, 2500k, and 3000k, and the number of subordinate blocks is 1000. Compare the query response time when Dabble method stores 10, 20, 30, 400, 500,000 pieces of data in the main block and 1000 pieces of data in the subordinate block. The experimental results are shown in Fig. 4.

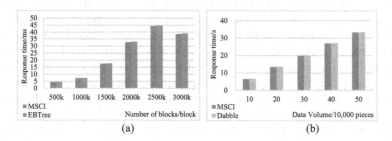

Fig. 4. Index query time comparison.

As can be seen from Fig. 4, compared with the existing methods, the index construction time of MSCI is optimized by about 7.44% and 5.89%, respectively, and when the data volume increases, the advantage of MSCI is more obvious than that of EBTree.

5.3 Comparison of Memory Consumption

The weight matrix constructed by MSCI in the slicing stage hardly takes up memory, and the master chain builds the index based on the jump consistent hash algorithm. Compared with the classical consistent hash, the jump consistent hash has almost no additional memory consumption, so the memory overhead in MSCI mainly considers the index construction of the slave blockchain. The false-positive of IBF is set to 0.0137 bits. The EBTree method rewrites the blockchain structure, and the memory consumption is mainly blocked data. Therefore, this section will compare the Dabble method. The experimental results are shown in Table 2.

Table 2. Memory Consumption Comparison.

Method	Memory Usage
Dabble	24 MB + 4 KB
MSCI	24 MB + 2.048 KB

As can be seen from Table 2, the Lognormal data set takes up 24MB of memory, while the neural network takes up 4KB of memory in Dabble method, while the IBF in MSCI still only takes up about 2.048KB of memory within the allowable range of false positives. Even if IBF and BF are in the same false positives, they can still keep the same order of magnitude.

6 Conclusion

With the wide application of blockchain technology, the traditional single-chain structure has gradually failed to meet the needs. Adding slave blockchain on the basis of the single-chain will make blockchain be applied to a wider range of fields, such as education chain and food chain. This paper proposes a composite index construction method for master-slave blockchain. Specifically, the whole master-slave blockchain structure is sliced according to the master chain, and the weight matrix is used to improve the maintainability of segmentation, which provides support for index construction. On this basis, aiming at the characteristics of different data scales between master chain and slave chain, the index construction method based on jump consistency hash algorithm and the index construction method of slave chain based on IBF is proposed to improve the query efficiency of the master-slave blockchain. Experimental results show that the proposed method has great advantages over the existing methods in terms of construction time, query efficiency, and memory consumption.

Acknowledgment. This study was supported by the Applied Basic Research Program of Liaoning Province (No. 2022JH2/101300250); the Digital Liaoning Smart Building Strong Province (Direction of Digital Economy) (No.13031307053000568); the National Key R&D Program of China (No. 2021YFF0901004); the Central Government Guides Local Science and Technology Development Foundation Project of Liaoning Province (No. 2022JH6/100100032); the Natural Science Foundation of Liaoning Province (No. 2022-KF-13-06).

References

1. Zhu, J.M., Zhagn, Q.N., Gao, S., Ding, Q.Y., Yuan, L.P., et al.: Blockchain-based trusted federated learning model for privacy protection. Chin. J. Comput. **44**(12), 2464–2484 (2021)
2. Li, C., Li, P., Zhou, D., et al.: A decentralized blockchain with high throughput and fast confirmation. In: 2020 {USENIX} Annual Technical Conference ({USENIX}{ATC} 2020), pp. 515–528 (2020)
3. Połap, D., Srivastava, G., Jolfaei, A., et al.: Blockchain technology and neural networks for the internet of medical things. In: IEEE INFOCOM 2020-IEEE Conference on Computer Communications Workshops (INFOCOM WKSHPS), pp. 508–513. IEEE (2020)
4. Wei, S.J., Li, S.S., Wanag, J.H.: Cross-domain authentication protocols based on identity cryptosystems and blockchains. Chin. J. Comput. **44**(05), 908–920 (2021)
5. Bao, J., He, D., Luo, M., et al.: A survey of blockchain applications in the energy sector. IEEE Syst. J. **15**(3), 3370–3381 (2020)
6. Kalodner, H., Möser, M., Lee, K., et al.: {BlockSci}: design and applications of a blockchain analysis platform. In: 29th USENIX Security Symposium (USENIX Security 2020), pp. 2721–2738 (2020)
7. Huang, J., Kong, L., Chen, G., et al.: Towards secure industrial IoT: blockchain system with credit-based consensus mechanism. IEEE Trans. Industr. Inf. **15**(6), 3680–3689 (2019)
8. Sui, Y., Wang, W., Deng, X.: High throughput verifiable query method for blockchain-oriented off-chain database. J. Chin. Comput. Syst. **42**(6), 1304–1312 (2021)
9. Cai, L., Zhu, Y.C., Guo, Q.X., Zhang, Z., Jin, C.Q.: Efficient materialized view maintenance and trusted query for blockchain. J. Softw. **31**(3), 680–694 (2020)
10. Nathan, V., Ding, J., Alizadeh, M., Kraska, T.: Learning multi-dimensional indexes. In: 2020 ACM SIGMOD International Conference on Management of Data (SIGMOD 2020), 14–19 June 2020, Portland, OR, USA, 16 p. ACM, New York (2020)
11. XiaoJu, H., XueQing, G., ZhiGang, H., et al.: Ebtree: a b-plus tree based index for Ethereum blockchain data. In: Proceedings of the 2020 Asia Service Sciences and Software Engineering Conference, pp. 83–90 (2020)
12. Kipf, A., Marcus, R., van Renen, A., et al.: RadixSpline: a single-pass learned index. In: Proceedings of the Third International Workshop on Exploiting Artificial Intelligence Techniques for Data Management, pp. 1–5 (2020)
13. Xing, X., Chen, Y., Li, T., et al.: A blockchain index structure based on subchain query. J. Cloud Comput. **10**(1), 1–11 (2021)
14. Alghamdi, N., Zhang, L., Zhang, H., et al.: ChainLink: indexing big time series data for long subsequence matching. In: 2020 IEEE 36th International Conference on Data Engineering (ICDE), pp. 529–540. IEEE (2020)
15. Gao, Y.N., Ye, J.B., Yang, N.Z., Gao, X.F., Chen, G.H.: Middle layer-based scalable learned index scheme. J. Softw. **31**(3), 620–633 (2020)

Multi-source Heterogeneous Blockchain Data Quality Assessment Model

Ran Zhang, Su Li, Junxiang Ding, Chuanbao Zhang, Likuan Du, and Junlu Wang[✉]

School of Information, Liaoning University, Shenyang 110036, China
wangjunlu@lnu.edu.cn

Abstract. Blockchain-based applications are becoming more and more widespread in business operations. This paper proposes a multi-source heterogeneous blockchain data quality assessment method for enterprise business activities, aiming at the problems that most of the data in enterprise business activities come from different data sources, information representation is inconsistent, information ambiguity between the same block chain is serious, and it is difficult to evaluate the consistency, credibility and value of information. The method realizes the consistency assessment by calculating the similarity of block information. After that, a trustworthiness characterisation method is proposed based on information sources and information comments, to obtain the trustworthiness assessment of the business. Finally, based on the information trustworthiness characterization, a value assessment method is introduced to assess the total value of business activity information in the blockchain, and a blockchain quality assessment model is constructed. The experimental results show that the proposed model has great advantages over existing methods in assessing inter-block consistency, intra-block activity information trustworthiness and the value of blockchain.

Keywords: Blockchain · Semantic similarity between blocks · Content credibility · Block value · Assessment model

1 Introduction

Blockchain [1–3] is a new distributed technology for generating, storing, manipulating and verifying data through block-chain structure, consensus algorithm and smart contract [4, 5]. It can achieve value communication between trustless [6, 7] nodes without relying on third-party trusted institutions. These characteristics make blockchain widely used in enterprise business activities, such as upstream and downstream supply chains, digital assets, business event monitoring, enterprise credit investigation, etc. Large domestic and foreign enterprise units, such as Google, Baidu and Alibaba, have established their own enterprise federated blockchain systems.

The quality of the information in existing enterprise blockchain systems varies. The information on business activities mostly originates from different fields and institutions, and the way of information representation is ambiguous. Moreover, it is difficult to ensure the credibility and value of the information in the block due to the constraints

S. Yang and S. Islam (Eds.): APWeb-WAIM 2022 Workshops, CCIS 1784, pp. 86–94, 2023.
https://doi.org/10.1007/978-981-99-1354-1_9

of the enterprise's own credibility and other conditions. Therefore, in the process of block establishment, there are problems of inconsistent information between blocks and inconsistent indicators such as credibility and value of content.

Moreover, traditional evaluation methods do not make full use of the features of blockchain of leaving traces throughout the process, being non-temperable and traceable, and the evaluation efficiency and accuracy are low, resulting in the inability of enterprise users and relevant regulatory authorities to quickly screen out the blockchain that meets their needs and establish a unified analysis model [8, 9]. Therefore, effective evaluation of enterprise blockchains is a hot spot and difficult area of research in the blockchain field.

To address these issues, this paper proposes a multi-source heterogeneous blockchain data quality assessment model for enterprise business activities to achieve an efficient assessment of the consistency, credibility and value of business activity information. The main contributions of this paper are as follows:

(1) To address the problem of inconsistent and poor accuracy of enterprise business activity information in the blockchain, a method to calculate semantic similarity between blocks is proposed to evaluate information consistency of block chain.
(2) On this basis, aiming at the difficulty of information credibility assessment of multi-source enterprises' business activities in blockchain, a method of information credibility characterization based on information source and evaluation is proposed, and the representation results of all parts are integrated.
(3) Aiming at the problem of value evaluation of active information in block chain, a method of information value evaluation is proposed to measure the value of business activities in block. Finally, by integrating semantic similarity, content assessment and value assessment between blocks, a multivariate heterogeneous blockchain data quality assessment model for enterprise business activities is obtained.

2   Related Studies

At present, many scholars have conducted in-depth research on data quality assessment methods and have achieved certain research results.

In blockchain information consistency evaluation, literature [10] proposed a structured gradient tree boosting (SGTB) algorithm for entity disambiguation method, which has better performance in cross-domain evaluation, but it ignores the role of contextual information of entities on the computation process; literature [11] proposed a factor graph-based inconsistent record pair computation method, which parses entities but does not correlate entity representation with the category it belongs to.

In terms of blockchain information credibility assessment, literature [12] proposes a data credibility assessment method based on multi-source heterogeneous information fusion, which shows better results in improving the convergence of credibility calculation, but ignores the information content itself credibility; literature [13] proposes a supervised machine learning method for user-generated content credibility assessment, but the method only focuses on relevant comment information, ignoring the attention to information content and its sources.

In terms of blockchain information value assessment, literature [14] proposes a confidence-based reliability assessment method, which can accurately obtain the distribution to which the data belongs, but ignores where the value of the data itself lies; literature [15] combines variable weight theory and cloud model theory to construct a VW&ICM computational model for risk assessment, which weakens the influence of subjective factors on the assessment results, but the model focuses more on the weight assignment factors and ignores the information of the overall value of the data.

In summary, this paper proposes a multi-source heterogeneous blockchain quality assessment model for enterprise business activities by addressing the shortcomings of existing blockchain enterprise business activity information assessment methods and considering the consistency, credibility and value of data.

3   Calculation of Semantic Similarity Between Blocks

As the information of business activities in blockchain mostly comes from different data sources, which leads to inconsistent information representation, such as data record format, entity and activity name designation methods. This leads to ambiguity and low data quality in the business activity information stored in the blockchain. To solve this problem, this paper evaluates the consistency of blockchain data by comparing semantic similarity between entities.

In the inter-block similarity measure, assuming that the total number of blocks in the blockchain is N, this paper takes the first block in the blockchain as the block to be compared and the other blocks as the candidate comparison blocks, and uses the average value of the similarity between the block to be compared and the other blocks as the consistency measure of this blockchain, i.e., the average similarity measure.

The semantic similarity is calculated using the name of the business entity and the specific content of the business transaction activities, using $SimText(A,B)$ to denote the semantic similarity of the blocks represented by block A and block B, respectively, and vectorizing the semantic description information represented by block A and block B as $A = \{m_{11}, m_{12}, m_{13}, ...\}$ and $B = \{m_{21}, m_{22}, m_{23}, ...\}$, respectively, cosine similarity calculation can be expressed as:

$$SimText(A, B) = \frac{A \cdot B}{\|A\| \|B\|} \tag{1}$$

Finally, the total consistency calculation of the blockchain can be expressed as the average of the similarity of these blocks, the average similarity measure:

$$CoHerence = \frac{\sum_{i=2}^{N} SimText(A, i)}{N} \tag{2}$$

where N is the total number of blocks to be compared, i is the blocks other than the blocks to be compared.

4 Blockchain Quality Assessment Model Construction

The evaluation model measures the consistency of the blockchain, the credibility of the information, and the value of the information, and assigns different weights according to the importance.

4.1 Blockchain Content Evaluation Based on Trustworthiness Theory

According to the characteristics and attributes of credibility, the information source, comment are used to independently represent the credibility of information, and the representation results are fused.

Source-Based Information Trustworthiness Characterization

(1) Credibility of information pages

The reliability of an information source is judged by the reachability of links in a page and the availability of pages reached. The specific model is as follows:

$$Page\_R(i_{A,B,C}) = \frac{A}{A \cup B} - \frac{C}{A} \tag{3}$$

where $Page\_R(i_{A,B,C})$ denotes the credibility of the information page, A denotes the set of reachable links among the relevant links, B denotes the set of unreachable links among the relevant links, and C denotes the set of unavailable pages among the pages reached by the links.

(2) Credibility of the information publisher

The average trust probability of publishers is calculated as follows:

$$Publisher\_R(i) = \frac{1}{N} \sum_{j=1}^{N} Publisher\_R_{j \to i}, i \neq j \tag{4}$$

$Publish\_R$ is the publisher's current credibility.

(3) The fusion of credibility characterization results of the source

The source-based credibility is calculated as follows:

$$R\_source(i) = \sqrt{\alpha Page\_R(i)^2 + \beta Publisher\_R(i)^2} \tag{5}$$

Information Trustworthiness Characterization Based on Comment Features

The analysis of representational tendency for comments is performed in two steps: first, the relevance of the comments to the message is calculated; second, the representational tendency value is calculated for the relevant comments.

1) Relevance of comments to information

The specific calculation of relevance between comments and information is as follows:

$$P(C|I) = \lambda \prod_{w \in C} p(w|I) + (1 - \lambda) \prod_{w \in C} \sum_{t \in t_I} \prod_{w \in t} p(w|t) * p(t|I) \quad (6)$$

where I is the message, C is an entry in the set of comments of I, w is a word in C, $p(C|I)$ is the probability of I generating C, and $p(w|I)$ is the probability of w appearing in I. t_I is the set of topics of message I, t is a topic in t_I, λ is a parameter, $p(w|t)$ is the probability of word w appearing in topic t, and $p(t|I)$ is the probability of topic t appearing in message I. $p(C|I)$ is calculated for all comments in message I.

2) Related commentary on representational tendencies

The representation propensity values were calculated as follows:

$$R\_review = \sum_i (1 + p(a_i)/count(a|R)) \times w(a_i) \quad (7)$$

Finally, information credibility in blockchain is a comprehensive measure of source-based credibility and evaluation-based credibility.

$$Accu = \alpha R\_source(i) + \beta R\_review \quad (8)$$

Among them, the credibility of information source accounts for the largest proportion, and the credibility based on evaluation accounts for a small proportion due to the subjectivity of evaluation.

4.2 Blockchain Value Assessment Based on the Amount of Information

To measure the validity of information on business activities in the blockchain, this paper adopts the method of calculating the amount of information, which is measured by the amount of information on business activities in the block. Each block can be regarded as a discrete source, and if it is a discrete random variable (a certain chain), the set of values of the random variable and its probability measure are as follows:

$$X = \{x_1, x_2, ..., x_n\}, p_i = P[X = x_i] \quad (9)$$

The amount of information can be defined as follows:

$$Validity = -\sum_{i=1}^{n} p(x) \log_2 p(x), p(x) = P(X = x), \quad (10)$$

Here, X is a random variable, $Validity$ is the amount of information contained in the block, and $p(x)$ is the probability distribution function of the variable X.

4.3 Quality Assessment Model

In this paper, we propose a weighted approach to construct a blockchain quality assessment model, which uses the similarity of block representations, source-based block trustworthiness representations, and the weighted average of the values contained in the blocks to measure the comprehensive quality of the blockchain. The final evaluation model is as follows:

$$Q = \alpha CoHerence + \beta Accu + \gamma Validity \tag{11}$$

Since blockchain value is an important indicator of blockchain quality, the greater the value contained in the blockchain, the greater the application value of that blockchain, followed by information trustworthiness, hence the weight $\gamma > \beta > \alpha$.

5 Experiments and Results

The experimental platform is an Intel Core i5-7400@3GHz processor with 8 GB processor memory and 64-bit Windows 10 operating system. The experimental dataset is searched on search engines with the subject term of business activities and crawled official data, about 110,000 items, and then simple data cleaning is performed by filling in missing values, smoothing or removing outlier points and other operations to get about 100,000 items of data, and the details of the dataset are listed in Table 1. In this paper, simulation experiments are conducted in terms of model evaluation efficiency, and the model built in this paper (DQAM) is used to compare with other models, and the comparison models are AHP, DSMM, etc.

Table 1. Experimental data sets

Entity Name	Bytes
name	16bytes
type	8 bytes
activityOntology	Variable length bytes
comment	Variable length bytes
informationSource	16 bytes

5.1 Accuracy Assessment Efficiency Comparison

This section validates the efficiency of the trustworthiness theory-based blockchain content evaluation (CEBT) method. The vertical axis is the operational efficiency, and the horizontal axis represents the number of data entries, in units of ten thousand, and the specific operational efficiency comparison of each method on different data sets is shown in Fig. 1. As can be seen from the figure, the average evaluation efficiency of the CEBT method is higher than that of the other methods.

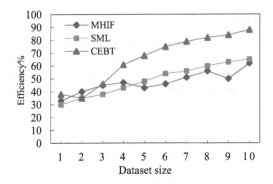

Fig. 1. Accuracy assessment efficiency comparison

5.2   Comparison of Value Assessment Efficiency

The experiments were conducted by simulating the comparison of the value assessment efficiency using the amount of information approach (VIS) with the value assessment approach based on data distribution (VADD) and the VW&ICM calculation model, with the horizontal coordinate indicating the amount of data, in units of ten thousand, and the vertical coordinate indicating the assessment efficiency, and the experimental results are shown in Fig. 2.

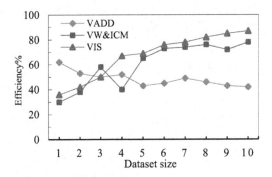

Fig. 2. Valuation efficiency comparison

From the experimental results, it can be found that the information amount based assessment method proposed in this paper is more efficient. The main reason for this is that the information-amount-based data value assessment focuses on the data value itself rather than its secondary factors such as distribution and regularity, so the method is intuitive, clear and more efficient.

5.3   Total Evaluation Efficiency Comparison

This section compares the running efficiency of each model, the horizontal coordinate represents the entity data set in the blockchain, in units of ten thousand, the vertical coordinate indicates the time required to run each model, and the experimental results are shown in Fig. 3.

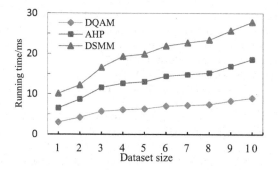

Fig. 3. Model efficiency comparison

6 Conclusion

This paper proposes a multi-source heterogeneous blockchain data quality assessment model for enterprise business activities. Firstly, the degree of semantic similarity between blocks is calculated. Then, based on the content assessment of blockchain, the credibility evaluation of blockchain information can be obtained through the credibility characterization method based on information source, information comment and business activity information content to fuse all characterization results. Finally, the value assessment method is used to evaluate the value amount contained in the content of the blockchain, and the above information is combined to obtain the blockchain quality assessment model.

Acknowledgement. This study was supported by the Applied Basic Research Program of Liaoning Province (No. 2022JH2/101300250); the Digital Liaoning Smart Building Strong Province (Direction of Digital Economy) (No. 13031307053000568); the National Key R&D Program of China (No. 2021YFF0901004); the Central Government Guides Local Science and Technology Development Foundation Project of Liaoning Province (No. 2022JH6/100100032); the Natural Science Foundation of Liaoning Province (No. 2022-KF-13-06).

References

1. He, X., Wang, J., Liu, J., et al.: Smart grid nontechnical loss detection based on power gateway consortium blockchain. Secur. Commun. Netw. **2021** (2021)
2. Shen, M., Sang, A.Q., Zhu, L.H., Sun, R.G., Zhang, C.: Recognition method of abnormal transaction behavior of blockchain digital currency based on motivation analysis. J. Comput. **44**(01), 193–208 (2021)
3. Hong, S.: Research on sharding model for enabling cross heterogeneous blockchain transactions. J. Digit. Converg. **19**(5), 315–320 (2021)

4. Fu, L.Q., Tian, H.B.: Ethereum voting protocol based on smart contract. J. Softw. **30**(11), 3486–3502 (2019)
5. Wang, X.B., Yang, X.Y., Shu, X.F., Zhao, L.: Formal verification of smart contract for MSVL. J. Softw. **32**(6), 1849–1866 (2021)
6. Truong, N., Lee, G.M., Sun, K., et al.: A blockchain-based trust system for decentralised applications: when trustless needs trust. Futur. Gener. Comput. Syst. **124**, 68–79 (2021)
7. Lee, G.M.: A blockchain-based trust system for decentralised applications: when trustless needs trust. Future Gener. Comput. Syst. **124**, 68–79 (2021)
8. Colomo-Palacios, R., Sánchez-Gordón, M., Arias-Aranda, D.: A critical review on blockchain assessment initiatives: a technology evolution viewpoint. J. Softw. Evol. Process **32**(11), e2272 (2020)
9. Zhang, A., Zhong, R.Y., Farooque, M., et al.: Blockchain-based life cycle assessment: an implementation framework and system architecture. Resour. Conserv. Recycl. **152**, 104512 (2020)
10. Yang, Y., Irsoy, O., Rahman, K.S.: Collective entity disambiguation with structured gradient tree boosting. arXiv preprint arXiv:1802.10229 (2018)
11. Xu, Y.L., Li, Z.H., Chen, Q., Wang, Y.Y., Fan, F.F.: Disambiguation method of inconsistent records based on factor graph. Comput. Res. Dev. **57**(01), 175–187 (2020)
12. Yanling, F., et al.: Credibility assessment method of sensor data based on multi-source heterogeneous information fusion. Sensors **21**(7), 2542 (2021)
13. Jain, P.K., Pamula, R., Ansari, S.: A supervised machine learning approach for the credibility assessment of user-generated content. Wireless Pers. Commun. **118**(4), 2469–2485 (2021)
14. Moon, M.Y., Cho, H., Choi, K.K., et al.: Confidence-based reliability assessment considering limited numbers of both input and output test data. Struct. Multidiscip. Optim. **57**(5), 2027–2043 (2018)
15. Lin, C., Zhang, M., Zhou, Z., et al.: A new quantitative method for risk assessment of water inrush in karst tunnels based on variable weight function and improved cloud model. Tunn. Undergr. Space Technol. **95**, 103136 (2020)

The Third International Workshop on Deep Learning in Large-Scale Unstructured Data Analytics

A Multimodal Activation Detection Model for Wake-Free Robots

Hangming Zhang[1], Jianming Wang[1], Shengjiao Yang[2], and Yukuan Sun[3(✉)]

[1] School of Computer Science and Technology, Tiangong University, Tianjin, China
{2031081014,wangjianming}@tiangong.edu.cn
[2] School of Psychology, Guizhou Normal University, Guiyang, China
19010250521@gznu.edu.cn
[3] Center for Engineering Internship and Training, Tiangong University, Tianjin, China
sunyukuan@tiangong.edu.cn

Abstract. During an interaction with a robot, the first thing we usually do is wake up the robot using a wake word. For example: 'XiaoduXiaodu' and 'Hey, Siri', these wake words undoubtedly reduce the interaction experience between us and robots. In this work, we focus on interacting with the robot without the use of wake words, even when the user is not within the robot's field of view. To accomplish this task, we propose a multimodal activation detection model (MADM), which consists of three parts: primary feature extraction, high-level feature fusion, and fused feature classification. The first part is used to extract the original video and audio as primary feature vectors. The second part uses our proposed local variable weight fusion strategy to convert primary features into high-level features and fuse them into fused features for classification. The three parts use a fully connected neural network to classify the fused features to determine whether a response is required. To evaluate MADM, we constructed a dataset containing 7992 short videos recorded by 99 invited volunteers. Extensive experiments demonstrate the effectiveness of our model and the necessity of a feature fusion strategy.

Keywords: Multimodal Activation · Wake Words · Deep Learning

1 Introduction

In the era of artificial intelligence, researchers have made great breakthroughs in the field of speech recognition [24], and due to the rapid progress of this technology, robots have become a necessity in people's daily lives. For example, people can control music playback, air conditioner switch, and set alarms through voice. People no longer need to study the functions of those cold buttons and only need a voice command to achieve the desired effect, thus bringing great

This work was supported by The National Natural Science Foundation of China (62072335) and The Tianjin Science and Technology Program (19PTZWHZ00020).

Fig. 1. Example of activating Xiaodu smart speakers using the wake word 'Xiaodu Xiaodu'.

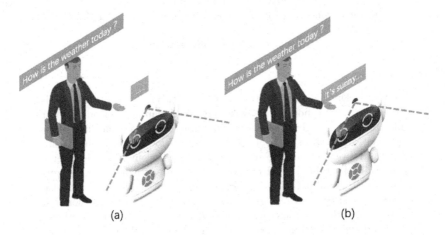

Fig. 2. An example of an interaction where the user is out of the robot's field of view. In Figure a, the user is standing behind the robot and interacting. Since the robot does not detect the wake word and does not see the user, the traditional method does not make the robot enter the active state. Figure b shows the ideal situation, where the robot switches to the active state to interact with the user.

convenience to people's daily life. According to a related report released by Strategy Analytics, following a 24% increase in the annual sales of conversational robots in 2020, the annual sales of conversational robots will accelerate by 31% in 2021, and this growth will continue in the next few years.

Before interacting with the robot, the user must use specific words to wake up. These specific words are called wake-up words, such as 'Hello Toyota' and 'Xiaodu Xiaodu'. Wake words are critical for when a conversational bot responds to a user's question and when it just listens and doesn't respond. Figure 1 shows an example of activating a dialogue bot using the wake-up word 'XiaoduXiaodu'. It can be seen that each time the user interacts with the robot, he needs to activate the robot with a wake-up word, even if the user is facing the robot. Although wake word technology has become a bridge for us to interact with robots, compared with real-life human-to-human communication, wake word technology will appear blunt and cumbersome. Especially in face-to-face interactions, an action and a smile are enough to express the intention of chatting.

Since the traditional multimodal activation detection model [14] can only achieve face-to-face activation detection, that is to say, the robot will switch to the activation state only when we are facing the robot and have the intention to talk to the robot. This greatly limits the interactive experience between the user and the robot, such as in Fig. 2(a), when the user stands behind the robot and says, 'What's the weather like today? ', the robot does not respond at this time, because no one appears in the robot's field of vision, and the robot will process this sentence as noise. Figure 2(b) is the ideal state, where the robot switches to the active state and answers the user's question.

To address this issue, we propose a Multimodal Activation Detection Model (MADM). We utilize two types of features, namely speech and video features, and use an efficient feature fusion strategy to fuse the two types of features into a vector for prediction. Our model is divided into three stages: primary feature extraction, advanced feature fusion, and fused feature classification. First, for the video modality, we use ResNet-152 [6] to extract its video features; for the audio modality, we extract the acoustic features, i.e. MFCC features, and then use CNN to project the MFCC features into the same space as the video features. Second, we use a local variable weight fusion strategy to fuse the two types of features and finally send the fused vector into the MLP for prediction. Our results show that multimodal information contributes more to model performance than unimodal information. Furthermore, our fusion strategy is more effective than the global variable weight fusion strategy and simply concatenates the two types of features.

Our main contributions are summarized as follows:

- We propose a novel multimodal activation detection model that explores the role of audio and video modalities in wake-free research. The purpose is to solve the problem that the existing activation detection model can only achieve face-to-face wake-free interaction, and the real wake-free interaction is that even if the human is not within the scope of the robot's camera, only one sentence can realize the wake-free interaction with the robot.
- Due to the inconsistency of multimodal information, we propose a local variable weight fusion strategy using fully connected neural networks. The policy can learn different weights for different types of data.
- Most importantly, we construct a new dataset with high-quality annotations. The dataset contains 7992 short videos recorded by 99 volunteers, filling the gap in the dataset created by Nie et al. [14] that only face-to-face interaction can activate.
- We conduct a series of experiments to demonstrate the effectiveness of our model. Our model achieves a 35.1% improvement in F1-score over the method using a baseline for activation detection.

2   Related Works

Our work involves wake word detection and other activation detection methods for robots.

2.1   Wake Word Detection

Vehicle-mounted interactive robots are ubiquitous in people's daily life. It can help drivers and passengers with many tasks, such as switching air conditioning, navigation, and more. This technology of detecting wake words from sound signals has been widely used to activate smart devices such as in-vehicle interactive robots and smart speakers. For example, the IMS intelligent interactive system carried by Changan Automobile will always be in a standby state. Only after hearing the predefined wake-up word "Xiao An, hello" will it enter a higher power consumption state to identify the commands used.

But wake word technology cannot distinguish whether a particular word is used for the wake or reference context. For example, the specific words in the sentence "Alexa, what time is it?" are arousal, while the specific words in the sentence "My friend Alexa likes to travel" are used for context. Although this problem can be solved by using words that are not common in daily life or more complex words, some scholars have devoted themselves to solving this problem theoretically. Wu et al. [21] introduced a two-stage wake word system based on deep neural network acoustic modelling and proposed a new method for modelling non-keyword contextual events using monophone-based units. Wang et al. [20] proposed a Streaming Transformers method to model wake word detection. Kumatani et al. [11] developed a technique to train features directly from single-channel speech waveforms to improve wake word detection performance. Jose et al. [9] proposed two new methods using single-stage word-level wake-word endpoint detection, which is the first study on single-stage wake-word endpoint detection. Hossain et al. [7] studied the design of an efficient corpus for wake word detection. Khursheed et al. [10] proposed a Tiny-CRNN (Tiny Convolutional Recurrent Neural Network) model for wake word detection, which has lower parameters and lower error rates than convolutional neural network models. Wang et al. [19] proposed a new method to train a hybrid DNN/HMM wake word detection system from partially trained labeled data, which enables online detection. Zhang et al. [23] proposed that the use of prosodic features helps to distinguish wake word sounds from other sounds.

Since wake word detection needs to keep the device in a listening state, it may lead to leakage of user privacy. Therefore, in order to solve the privacy problem, some scholars [2,3,18] use ultrasonic signals to prevent the speaker from recording, Iravantchi et al. [8] proposes the use of inaudible frequencies for privacy-preserving recordings. Feng et al. [5] provided a continuous authentication system and Coucke et al. [4] introduced a platform for processing speech on a microprocessor.

2.2   Other Activation Detection Methods

The activation method of wake word detection has many drawbacks, so many scholars have proposed other activation detection methods. Abraham et al. [12] proposed to use two cues, gaze direction and speech volume level, as switches for device activation. Yang et al. [22] and Ahuja et al. [1] explored using voice only to infer the user's spatial position and head orientation, and then used microphone arrays to design voice interactions. Pomykalski et al. [16] proposed the use of gesture triggers to activate dialogue bots, Yue et al. [17] proposed to use convolutional neural networks to detect sounds close to the microphone, that is, the characteristics of the popping sound when the user blows into the microphone when speaking, as a sign of device activation. Recently, Momeni et al. [13] proposed a novel convolutional architecture that can locate keywords based on the user's face, while audio tracks can be used to boost performance. Nie et al. [14] constructed the first multimodal activation detection dataset and proposed a method to solve the multimodal activation detection task using audio-visual consistency detection and semantic intent inference, which allows device-oriented of users do not need a wake word to activate the device. But in most situations in daily life, people do not always maintain a face-to-face state when interacting with robots. On the contrary, people prefer to treat robots as friends.

Therefore, in this paper, in addition to using multiple modalities, we focus more on solving dialogue scenarios where the user is not within the robot's field of vision. In addition, we also focus on better fusion of multimodal features.

3   Our Method

Our network architecture is shown in Fig. 3 and can be divided into three parts: (a) primary feature extraction, (b) advanced feature fusion, and (c) fusion feature classification.

3.1   Primary Feature Extraction

Video Features: Since the robot's vision can not only see the user, but also see the items carried by the user's body parts, the video features are mainly used to assist the network to detect whether there is a user within the robot's vision or whether the items carried by the user is related to the words spoken by the user. To obtain a visual representation of each speech, we first preprocess each frame by normalization, and then redefine the last layer in the ResNet-152 [6] image classification model pretrained by ImageNet to be 2048 dimensional, and use this model to extract the video features u_v of each video. It is worth noting that since the total number of frames F of each video is different, we calculate the average value of the $d_v = 2048$-dimensional feature vector u_i^v obtained for each frame as the video feature of each video.

$$u_v = \frac{1}{F} \left(\sum_i u_i^v \right) \in \mathbb{R}^{d_v} \tag{1}$$

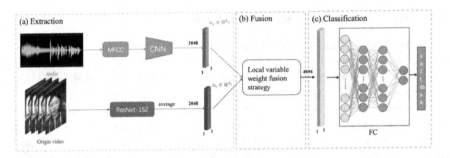

Fig. 3. The overall structure of our proposed model. It consists of three key steps: primary feature extraction, advanced feature fusion, and fused feature classification. Where MFCC represents the MFCC features extracted using the interface function provided by the python_speech_features toolkit, and CNN represents the single-layer CNN network proposed by Nie et al. [14] that maps MFCC features to the same dimension of video features. The local variable weight feature fusion strategy is described in detail in Sect. 3.2. FC stands for Fully Connected Neural Network and is described in detail in Sect. 3.3 (https://github.com/jameslyons/python_speech_features.)

Speech Features: To encode speech modalities, we use MFCC features with 13 cepstral numbers as primary features for speech modalities. MFCC features are considered to be a powerful audio descriptor in many speech processing tasks. To better capture the acoustic information relevant to activation detection, we segment the speech signal into multiple segments of 25 ms with a stride of 10 ms, where adjacent speech segments share some overlap due to edge effects of the window function. Since all speech signals are no longer than 15 s, we unify the length of all speech signal sequences to 1500, and perform zero-padded operations on shorter speech signals. The MFCC feature u_i^{MFCC} of the final audio modality can be expressed as $u_i^{MFCC} \in \mathbb{R}^{1500*13}$.

For the extracted MFCC features, we also use CNN for further processing. CNN has shown good performance in various speech processing tasks. Inspire by Nie et al. [14], we use a single-layer CNN to convert MFCC features to the same dimension $d_a = 2048$ as video features, Formally, We have

$$u_a = CNN(u_i^{MFCC}|\Phi_a) \in \mathbb{R}^{d_a} \tag{2}$$

where u_a represents the feature vector of the audio modality and Φ_a is the parameter of the CNN network. The parameter settings of the CNN network are the same as those of Nie et al. [14].

3.2 Local Variable Weight Fusion Strategy

In order to better fuse the features of the two modalities, we propose a local variable weight fusion strategy, as shown in Fig. 4. The fusion strategy can calculate different weights according to different samples. We first calculate the weight of the input feature through a multi-layer perceptual product and Softmax layer, so

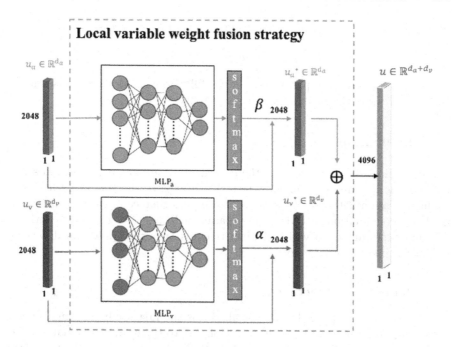

Fig. 4. Local variable weight fusion strategy graph, where α and β are the weights calculated by the multi-layer perceptual product MLP_v and MLP_a, $\oplus$ represents the connection operation, and u_a^* and u_v^* are the advanced features upgraded from the initial features.

as to upgrade the input primary feature to an advanced feature. Finally, all the transformed high-level features are connected together to form a fusion feature vector u.

$$\hat{\alpha} = Softmax(MLP_v(u_v|\Lambda_v)) \tag{3}$$

$$\hat{\beta} = Softmax(MLP_a(u_a|\Lambda_a)) \tag{4}$$

$$u = \hat{\alpha}u_v \oplus \hat{\beta}u_a \tag{5}$$

among them, Λ_v and Λ_a represent the parameters of the multi-layer perceptual product MLP_v and MLP_a, and these parameters can be learned and updated according to the gradient. α and β represent the calculated weights. $\oplus$ is the concatenation operation, which is used to concatenate high-level features for prediction.

3.3   Fusion Feature Classification

In this section, we input the fused feature vector u into a fully connected neural network, as shown in Fig. 3. Then go through a Softmax layer to get the final predicted activation probability. The activation function of the fully connected neural network is LeakyReLU, and the loss function uses cross-entropy.

$$\hat{y}_i = Softmax(FC(u|\Lambda_e)) \tag{6}$$

where Λ_e represents the parameters of the fully connected neural network FC, and y_i represents the final activation probability.

4 Experiment

In this section, we first introduce the dataset construction, experimental setup, baseline model, and experimental results. To explore the role of each modality in activation detection, we evaluated the performance of each modality and modality combination in the model separately. Finally, we conduct a series of ablation experiments to verify the effectiveness of our proposed local variable weight feature fusion strategy.

4.1 Dataset Construction

To the best of our knowledge, Nie et al. [14] is the first study on a multimodal activation task and created a dataset belonging to them. Since the dataset created by Nie et al. [14] is not public and no public dataset is available, to address this issue, we created our own dataset containing 7992 videos recorded by 99 invited volunteers. We focus on addressing dialogue scenarios where the user is out of the robot's field of view, so we collect eight types of video samples. As follows.

(1) The user is facing the camera of the robot and expresses a sentence with a dialogue intention, such as 'What time is it?';
(2) The user is not facing the camera of the robot, such as looking sideways, expressing sentences with dialogue intentions;
(3) The user is not in the field of vision of the robot, and the sound comes from the user or noise outside the camera, expressing a statement with a dialogue intention;
(4) The user is facing the camera of the robot, but the user in front of the camera does not make any sound, the sound comes from the user outside the camera or noise and the sentence expressed has no dialogue intention;
(5) The user is facing the camera of the robot, but the sentence expressed has no dialogue intention, such as "practice is the only criterion for testing truth";
(6) The user is not facing the camera of the robot, expressing a sentence without dialogue intention;
(7) The user is not in the field of view of the robot, the sound comes from the user or noise outside the camera, and the sentence expressed has no dialogue intention;
(8) The user is facing the camera of the robot, and the sentence expressed has the intention of dialogue, but requires a handheld communication device, such as the scene of making a phone call with a friend.

Among them, the first three types of video tags are Positive, and the remaining types of video tags are Negative. Each type of video recorded by each volunteer is saved in a long video, after which we split the long video into several short videos, each of which contains the completed representation. In addition, we divide the data into training, validation and test sets with a ratio of 80%:10%:10%, and guarantee that people who appear in the training set will not appear in the validation or test set. To more accurately evaluate the model, we manually inspect the dataset to ensure completeness of representation and accuracy of labels. The statistics of our final dataset are shown in Table 1.

Table 1. Statistics for our dataset type.

Data type	Positive	Negative	Sum
Train	2477	4085	6562
Val	267	466	733
Test	282	415	697
Sum	3026	4966	7992

4.2   Experimental Setup

Our models are built on PyTorch 1.7.1 [15] and run on an NVIDIA GeForce RTX 3090 GPU. We use Adam as our optimizer and set the initial learning rate to 1e-5. The parameters of the ResNet-152 [6] model are not updated during training. Also, our task is actually a binary classification problem, so we use the widely used precision, recall and F1-score to evaluate the model. All hyperparameters of the experiment are listed in Table 2.

Table 2. Hyper-parameters table.

Hyper-parameters	Value
Learning rate	1e–5
Early stop patience	7
Dropout rate	0.5
Number of iterations	2000
MLP input size	2048
MLP hidden size	512
MLP output size	2
FC input size	4096
FC hidden size	512
FC output size	2

4.3   Baseline Model

The experiments are mainly conducted using two baseline methods:

Random: This baseline will randomly classify test samples.

SVM: This baseline will use a support vector machine (SVM) with an RBF kernel as the classifier for the model. SVM is a supervised learning method and is widely used in statistical classification and regression analysis. We employ a grid search strategy to set the optimal parameters for the kernel function coefficient gamma and penalty factor C for each experiment.

4.4   Experimental Results

Table 3. Performance comparison between different models and different modalities. The best results have been shown in bold, with A for speech modality and V for video modality.

Model	Modality	Precision	Recall	F1-score
Random	A+V	38.8	46.8	42.4
SVM	A	45.5	21.6	29.3
	V	62.7	22.7	33.3
	A+V	56.1	37.2	44.8
Our	A	**84.2**	66	74
	V	53.3	40.4	46
	A+V	83.4	**76.6**	**79.9**

To further explore the influence of two modalities on our proposed model, we used all possible inputs to evaluate our model: unimodal (A, V) and bimodal (A+V). The performance is shown in Table 3.

As can be seen from Table 3, the model performance (33.3% and 46%) obtained using only video features (V) is less than 50%, which indicates that using video feature information alone is not helpful for activation detection. On the other hand, the model performances (29.3% and 74%) obtained by the two methods using only audio features (A) are quite different, indicating that the SVM method is not suitable for classifying audio features. Besides, both the SVM classifier-based method used and our method, the performance of using bimodal features (A+V) is higher than that of using single-model features (A, V). This shows that our feature fusion strategy fuses the features of the two modalities well, and the single-modal features with low model performance do not lower the performance of the model using dual-model features.

4.5   Ablation Study

In this section, to evaluate our local variable weight fusion strategy, we first remove the feature fusion strategy, directly concatenate the features of the two modalities and then input them into the MLP for classification, resulting in a model(w/o F). Then we use a global weight fusion strategy with the following formula:

$$u = \gamma u_v \oplus \delta u_a \tag{7}$$

where γ and δ represent global weight parameters, and $\oplus$ represents the connection operation. In this ablation experiment, the initial values of γ and δ are both set to 1.

Table 4. Experimental results in an ablation study.

Model	Precision	Recall	F1-score
Model(w/o F)	79.8	**77**	78.3
Model(GFS)	81.3	**77**	79.1
Ours	**83.4**	76.6	**79.9**

The model (GFS) is obtained by training the global weight fusion strategy, and the model performance is shown in Table 4. It can be seen that the use of the global weight fusion strategy reduces the F1-score, but compared with the features of the two modalities directly concatenated, the F1-score is improved by 0.8% (from 78.3% to 79.1%), which indicates that using the string The fusion strategy of connected features will degrade the model performance even more. Compared with our proposed local variable weight fusion strategy, the global weight fusion strategy results in a 0.8% (from 79.9% to 79.1%) drop, which indicates that our proposed feature fusion strategy is necessary and can better integrate two modal characteristics.

5   Conclusion

In this paper, we propose a new model for multimodal activation detection. The model makes full use of information from both speech and video modalities to solve challenging multimodal activation tasks. In particular, our model can respond to scenarios where the user is not within the robot's field of view. Furthermore, the evaluation results demonstrate the effectiveness of our proposed model and the usefulness of the two modalities and the necessity of a local variable weight feature fusion strategy. In future work, we plan to deal with scenarios where more people appear in the robot's field of vision.

References

1. Ahuja, K., Kong, A., Goel, M., Harrison, C.: Direction-of-voice (dov) estimation for intuitive speech interaction with smart devices ecosystems. In: UIST, pp. 1121–1131 (2020)

2. Chandrasekaran, V., Banerjee, S., Mutlu, B., Fawaz, K.: Powercut and obfuscator: an exploration of the design space for privacy-preserving interventions for voice assistants. arXiv preprint arXiv:1812.00263 (2018)
3. Chen, Y., et al.: Wearable microphone jamming. In: Proceedings of the 2020 Chi Conference on Human Factors in Computing Systems, pp. 1–12 (2020)
4. Coucke, A., et al.: Snips voice platform: an embedded spoken language understanding system for private-by-design voice interfaces. arXiv preprint arXiv:1805.10190 (2018)
5. Feng, H., Fawaz, K., Shin, K.G.: Continuous authentication for voice assistants. In: Proceedings of the 23rd Annual International Conference on Mobile Computing and Networking, pp. 343–355 (2017)
6. He, K., Zhang, X., Ren, S., Sun, J.: Deep residual learning for image recognition. In: Proceedings of the IEEE Conference on Computer Vision and Pattern Recognition, pp. 770–778 (2016)
7. Hossain, D., Sato, Y.: Efficient corpus design for wake-word detection. In: 2021 IEEE Spoken Language Technology Workshop (SLT), pp. 1094–1100. IEEE (2021)
8. Iravantchi, Y., Ahuja, K., Goel, M., Harrison, C., Sample, A.: Privacymic: utilizing inaudible frequencies for privacy preserving daily activity recognition. In: Proceedings of the 2021 CHI Conference on Human Factors in Computing Systems, pp. 1–13 (2021)
9. Jose, C., Mishchenko, Y., Senechal, T., Shah, A., Escott, A., Vitaladevuni, S.: Accurate detection of wake word start and end using a CNN. arXiv preprint arXiv:2008.03790 (2020)
10. Khursheed, M.O., Jose, C., Kumar, R., Fu, G., Kulis, B., Cheekatmalla, S.K.: Tiny-crnn: streaming wakeword detection in a low footprint setting. In: 2021 IEEE Automatic Speech Recognition and Understanding Workshop (ASRU), pp. 541–547. IEEE (2021)
11. Kumatani, K., Panchapagesan, S., Wu, M., Kim, M., Strom, N., Tiwari, G., Mandai, A.: Direct modeling of raw audio with dnns for wake word detection. In: 2017 IEEE Automatic Speech Recognition and Understanding Workshop (ASRU). pp. 252–257. IEEE (2017)
12. Mhaidli, A.H., Venkatesh, M.K., Zou, Y., Schaub, F.: Listen only when spoken to: Interpersonal communication cues as smart speaker privacy controls. Proc. Priv. Enhancing Technol. **2020**(2), 251–270 (2020)
13. Momeni, L., Afouras, T., Stafylakis, T., Albanie, S., Zisserman, A.: Seeing wake words: audio-visual keyword spotting. arXiv preprint arXiv:2009.01225 (2020)
14. Nie, L., Jia, M., Song, X., Wu, G., Cheng, H., Gu, J.: Multimodal activation: awakening dialog robots without wake words. In: Proceedings of the 44th International ACM SIGIR Conference on Research and Development in Information Retrieval, pp. 491–500 (2021)
15. Paszke, A., et al.: Pytorch: an imperative style, high-performance deep learning library. In: Advances in Neural Information Processing Systems, vol. 32 (2019)
16. Pomykalski, P., Woźniak, M.P., Woźniak, P.W., Grudzień, K., Zhao, S., Romanowski, A.: Considering wake gestures for smart assistant use. In: Extended Abstracts of the 2020 CHI Conference on Human Factors in Computing Systems, pp. 1–8 (2020)
17. Qin, Y., Yu, C., Li, Z., Zhong, M., Yan, Y., Shi, Y.: Proximic: convenient voice activation via close-to-mic speech detected by a single microphone. In: Proceedings of the 2021 CHI Conference on Human Factors in Computing Systems, pp. 1–12 (2021)

18. Roy, N., Shen, S., Hassanieh, H., Choudhury, R.R.: Inaudible voice commands: the {Long-Range} attack and defense. In: 15th USENIX Symposium on Networked Systems Design and Implementation (NSDI 18), pp. 547–560 (2018)
19. Wang, Y., Lv, H., Povey, D., Xie, L., Khudanpur, S.: Wake word detection with alignment-free lattice-free mmi. arXiv preprint arXiv:2005.08347 (2020)
20. Wang, Y., Lv, H., Povey, D., Xie, L., Khudanpur, S.: Wake word detection with streaming transformers. In: ICASSP 2021–2021 IEEE International Conference on Acoustics, Speech and Signal Processing (ICASSP), pp. 5864–5868. IEEE (2021)
21. Wu, M., et al.: Monophone-based background modeling for two-stage on-device wake word detection. In: 2018 IEEE International Conference on Acoustics, Speech and Signal Processing (ICASSP), pp. 5494–5498. IEEE (2018)
22. Yang, J., Banerjee, G., Gupta, V., Lam, M.S., Landay, J.A.: Soundr: head position and orientation prediction using a microphone array. In: Proceedings of the 2020 CHI Conference on Human Factors in Computing Systems, pp. 1–12 (2020)
23. Zhang, X., Su, Z., Rekimoto, J.: Aware: intuitive device activation using prosody for natural voice interactions. In: CHI Conference on Human Factors in Computing Systems, pp. 1–16 (2022)
24. Zhang, Y., et al.: Bigssl: exploring the frontier of large-scale semi-supervised learning for automatic speech recognition. IEEE J. Sel. Topics Signal Process. **16**(6), 1519–1532 (2022)

Analysis of Public Opinion Evolution in Public Health Emergencies Based on Multi-fusion Model

Bin Zhang, Ximin Sun$^{(\boxtimes)}$, Jing Zhou, Xiaoming Li, Dan Liu, and Shuai Wang

State Grid Ecommerce Technology Co., Ltd, Beijing, China
1464932411@qq.com

Abstract. This paper analyzes the public opinion topics and the emotional fluctuations of netizens during the closure of the city due to the new crown epidemic, and reveals the correlation between public emotional fluctuations and public opinion topics during various periods of public health emergencies. We use the BERT-BiLSTM fusion model to efficiently capture the two-way relationship in the sentence and improve the accuracy of sentiment classification of Weibo text at the same time, to extract the hot topic feature words at different periods of the city closure by LDA mode; finally, the SEI7R model is used to simulate the prevention and control recommendations of various periods of public opinion proposed in this paper to verify the effectiveness of the prevention and control recommendations. The experimental results show that: The F value of the BERT-BiLSTM fusion model in the classification of public sentiment polarity can reach up to 0.907, which can effectively classify the sentiment of netizens; The LDA model can effectively dig out over time The theme characteristics of the text that have gradually evolved over time. The simulation results of the SEI7R model show that the countermeasures and suggestions put forward in this paper can effectively provide a theoretical basis and methodological reference for public opinion management of public health emergencies.

Keywords: Public emergencies · Sentiment analysis · BERT-BiLSTM · LDA

1 Introduction

Public health emergencies have the characteristics of diversity of causes, differences in distribution and complexity of harm, and the wide spread of public health and property impacts on the public. At the end of 2019, a major public health event broke out in my country, the new crown pneumonia epidemic (COVID-19, hereinafter referred to as the epidemic), which had a serious impact on the daily life and physical and mental health of the Chinese public. Since mid-to-late 2020, domestic epidemics have appeared in Guangzhou, Beijing, Nanjing, Inner Mongolia and other cities, showing regional characteristics and seriously hindering social and economic development. In recent years, with the continuous development of social media, new channels have been provided for the rapid development of online public opinion. Among them, social media represented

S. Yang and S. Islam (Eds.): APWeb-WAIM 2022 Workshops, CCIS 1784, pp. 110–124, 2023.
https://doi.org/10.1007/978-981-99-1354-1_11

by Weibo has become an important platform for netizens to express their emotions. According to the "2020 Weibo User Development Report", the average daily epidemic information on Weibo has reached 16.1 billion times. During the epidemic, the Weibo public discussion platform has prominent attributes and is the main platform for obtaining epidemic information and expressing one's emotions. The emotions, opinions, and opinions expressed by the public through Weibo gather and ferment to form public opinions and spread rapidly. The key events that occurred during the epidemic and the promulgated prevention and control policies and measures will attract a lot of attention and heated discussions among netizens. Public opinion often spreads rapidly on the Internet, affecting netizens' judgment on the epidemic and further exacerbating public panic. The influence of network public opinion on government decision-making and social stability is increasing, and the timely response of the government plays an important role in restraining the development of public opinion [1]. By analyzing online public opinion topics and corresponding netizens' emotions, judging the public's actual needs, event concerns and emotional changes, it can help relevant government departments to predict and control emergencies, and then make quick and timely responses.

2   Related Work

Internet public opinion refers to the popular Internet public opinion on the Internet with different views on social issues. It is a form of social public opinion [2], tendentious remarks and opinions. Against the background of the epidemic, the public faced with prevention and control measures such as city closure, isolation treatment, and home isolation, and negative emotions such as anxiety and panic began to spread, resulting in significant negative psychology [3]. In order to grasp the characteristics and laws of the public's emotional evolution in public opinion events, many researchers have carried out related research on the dissemination of network public opinion and the control measures [4]. Liu et al. [5] used Weibo comments as the data source, and adopted the Bi-LSTM model based on the attention mechanism to analyze the evolution of public sentiment; Cao et al. [6] topic information and explore the law of evolution of Weibo public opinion; Wang Dan et al. [7] used the super network and dynamic network analysis method to determine the key nodes of public opinion, and through the introduction of the thought leading mechanism to judge the emotional tendency of the key nodes of public opinion, and then provide different guidance Strategy. Chang et al. [8] used Weibo data during the epidemic to analyze the public's anxiety fluctuations and spatial and temporal differences, proving that the development of the epidemic will trigger the public's anxiety, which will continue to be amplified with the development of the epidemic. Wang et al. [9] constructed an SCNDR system dynamics model based on the SIR model, which effectively modeled the spread of Internet rumors in emergencies, effectively controlled the spread of Internet rumors in public opinion, and reduced the negative impact of rumors.

In terms of the research methods of text emotion analysis, scholars usually classify emotions based on emotion dictionary, machine learning and deep learning. Based on the classification method of emotion dictionary, that is, emotional tendency analysis is carried out through words with emotional color provided by emotion dictionary. Representative emotion dictionaries such as foreign SentiWordNet [10], General Inquirer

[11], Opinion Lexicon [12], Domestic HowNet [13], DUTIR Emotion Words Ontology Library [14] and NTUSD Emotion Dictionary [15]. Whisner et al. [16] used the SentiWordNet sentiment dictionary to complete the sentiment analysis of Twitter text, and combined the sentiment value to obtain the final sentiment score; Xiong et al. [17] based on HowNet word similarity, sentences based on semantic and grammatical distance The emotion judgment method improves the accuracy of the emotion judgment result; He [18] realizes the emotion preference judgment of blog text by introducing the emotion dictionary based on the text analysis method of semantic understanding. Sentiment classification based on machine learning usually adopts models such as Support Vector Machine (SVM), Maximum Entropy (ME) and Naive Bayesian Model (NBM). Pang uses support vector machines, naive Bayes and maximum entropy classifiers to classify text sentiments. The study found that support vector machines have the highest classification accuracy among the three [19]. Bi et al. [20] analyzed the sentiment tendency of the user's comment text based on the text feature words and the sentiment classifier of the Naive Bayes algorithm. With the continuous development of computer technology, text sentiment analysis methods based on deep learning have become a hot research direction in the field of classification. Scholars began to use Recurrent Neural Network (RNN), Convolutional Neural Networks (CNN), Long-Short Term Memory (LSTM) and Bidirectional Long-Short-Term Memory (Bidirectional Long-Short-Term Memory) networks. -Short Term Memory, Bi-LSTM) and other deep learning models are applied to the work of text sentiment classification. [21] synthesized the word vectors obtained by the CNN model into sentence vectors, and used the sentence vectors as features to train a multi-label classifier, which greatly improved the performance of sentiment classification. Hu et al. [22] constructed a deep attention LSTM model fused with topic features. Compared with the previous attention-based sentiment analysis model, the model has further improved the accuracy. Wu et al. [23] combined the LSTM model with the cognitive emotion evaluation model, and used the cognitive emotion evaluation model to label microblog texts with emotion, so as to train the LSTM model. The empirical results show that the classification accuracy of the model is higher than that of the traditional model. Machine Learning Classification Methods.

In view of the above, this article uses the improved BERT-BiLSTM model to classify the sentiments of netizens by analyzing the relevant comments of Weibo netizens on the closure of the city from January 23, 2020, to the end of the city closure in Wuhan on April 8, 2020, and use the DUTIR sentiment vocabulary ontology database to classify the emotions of netizens in a fine-grained manner, and analyze the emotional changes of netizens in various periods of the city closure. At the same time, the topic extraction model of LDA is used to study the topics of each period, to clarify the characteristics of the topics in each period and the relationship between them and public sentiment fluctuations. According to the results of sentiment analysis and topic feature extraction, corresponding public opinion management and control countermeasures are put forward. On the one hand, it objectively shows the emotional fluctuations of netizens when they face key events during the epidemic. It provides theoretical reference for guiding the emotions of netizens and shifting the focus of netizens, and provides decision support for the government to adopt emergency policies in response to health emergencies in the future.

3   Data Sources and Research Methods

3.1   Data Sources

With the development of new media, Weibo has become an important channel for public opinion dissemination of emergencies. The openness, dissemination and immediacy of Weibo make public communication break through the limitation of time and space, and the analysis of Weibo content can effectively reveal the relationship between relevant prevention and control measures and netizens' emotional tendencies and public opinion events. Affected by the epidemic on January 23, 2020, Wuhan City implemented the "city closure" policy until the "unblocking" policy was issued on April 8, 2020. During this period, a large number of netizens expressed their views and opinions on the Weibo platform. Therefore, this paper collects 300,000 Weibo comments related to emergency policies during the epidemic from January 23, 2019 to April 30, 2020. Preprocessed with word and part-of-speech tagging, and removed URL links, punctuation marks, special characters and microblog content irrelevant to keywords in the original text, a new corpus was obtained.

3.2   Research Methods

Research Framework. The framework of the netizen sentiment analysis model constructed in this study is shown in Fig. 1, which consists of three parts: 1) Preprocessing of the original microblog data. Construction of BERT-BiLSTM model and LDA model and analysis of sentiment and topic of public opinion events; 2) Proposition and verification of public opinion supervision and response mechanism. Firstly, the sentiments of netizens are classified through DU-TIR, the emotional vocabulary ontology database of Dalian University of Technology; secondly, word vectors are generated by the BERT model and input to the BiLSTM model to perform sentiment analysis on the corpus to obtain the sentiment distribution of netizens; at the same time, the LDA document topic generation model is introduced. The topic feature words of the text corpus are extracted, and the topic features of the text in each period of the epidemic are obtained. Finally, according to the emotional tendencies and topic characteristics of netizens, the emotional fluctuations of netizens during the epidemic are analyzed, focusing on the key points and popular topics that cause emotional fluctuations. Realize the advance supervision of public opinion events and take corresponding measures to guide public opinion. 3) Through the analysis of the public's emotional evolution and topic characteristics in the epidemic, corresponding control measures are proposed for different periods of public opinion development, and the regional epidemic event - the Weibo comments of netizens in the Shanghai tour group is used as the verification data set. The improved SEI7R model is used to simulate the proposed countermeasures and suggestions to verify the effectiveness of the suggestions proposed in this paper.

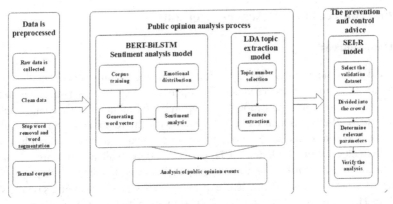

Fig. 1. Research framework

Sentiment Value Calculation. This paper takes the preprocessed text corpus as the object, integrates BERT model and BiLSTM model, and constructs a deep neural network model Bert-BiLSTM for sentiment classification in the field of public opinion. Firstly, the pre-processed text corpus is trained by BERT model and word vector is generated. Secondly, the word embedding vector output by BERT model is used as the input of BiLSTM model for feature extraction, so as to realize the emotion classification of corpus set. Finally, in order to realize the in-depth analysis of netizens' emotions in public opinion events, Based on the emotion vocabulary ontology database of Dalian University of Technology, the research divides netizens' emotions into seven categories.

Evaluation Indicators. This paper takes the preprocessed text corpus as the object, and integrates the BERT model with the BiLSTM model to construct a deep neural network model BERT-BiLSTM for sentiment classification in the field of public opinion. The model firstly trains the preprocessed text corpus by the BERT model to generate word vectors, and then uses the word embedding vector output by the BERT model as the input of the BiLSTM model for feature extraction, thereby realizing the sentiment classification of the corpus. Finally, in order to achieve for the in-depth analysis of netizens' emotions in public opinion events, the study divides netizens' emotions into seven categories according to the emotional vocabulary ontology database of Dalian University of Technology.

Evaluation Indicators. In this paper, indicators such as Precision, Recall, and F1 value are selected to evaluate the proposed model. The precision rate is the ratio of the number of correctly identified emotions of netizens to the number of recognized emotions of netizens, and the recall rate is the ratio of the number of correctly identified emotions of netizens to the total number of real emotions of netizens. F1 combines the values of precision rate and recall rate at the same time, so that both of them reach the highest value at the same time and achieve a balance. The specific formula is as follows.

$$P = \frac{\text{Correctly identify the number of emotions}}{\text{Number of emotions that have been identified}} \tag{1}$$

$$R = \frac{\text{Correctly identify the number of emotions}}{\text{The total number of correct emotions}} \tag{2}$$

$$F1 = \frac{2 \times P \times R}{P + R} \tag{3}$$

3.3 Model Accuracy

During the training of the BERT-BiLSTM model, 80% and 20% of the text corpus were randomly selected as the training set and validation set of the experiment. The specific parameters of the model are set as follows: the batch parameter train_bath_size is set to 128, the model to prevent overfitting dropout parameter is set to 0.5, the learning rate (learning_rate) is set to 0.0001, the max_length parameter is set to 100, and the epoch is set to 50. If the F1 value does not drop, the training is stopped, and the model with the highest F1 value is selected to perform sentiment analysis on the text corpus. In the end, the accuracy, recall and F1 value of the model in each sentiment analysis reached more than 80%, and the highest F1 value reached more than 90%. The experimental results of the model on text corpus classification are shown in Table 1.

Table 1. Classification results of text corpus

Emotional categories	Precision	Recall	F1
Anger	0.843	0.839	0.841
Hate	0.815	0.811	0.813
Fear	0.856	0.861	0.858
Sad	0.822	0.834	0.828
Surprised	0.806	0.828	0.817
Happy	0.904	0.911	0.907
Optimistic	0.826	0.841	0.833

4 Experimental Results

4.1 Topic Analysis

LDA is a document topic generation model, including word, topic and document three-layer structure. As a text mining technology for unsupervised machine learning, its three-layer Bayesian probability model can accurately identify hidden topics in Weibo comments. The LDA model believes that each word in a document is obtained through a process of "selecting a topic with a certain probability, and selecting a word from this topic with a certain probability". Before topic mining is performed on the text corpus of each period, the number of topics in each period text needs to be determined. Too many or too few topics, respectively, can lead to problems with too scattered topics and an inability to explain the details of the topic's content. Therefore, the selection of the number of topics has a key impact on the results of topic mining. Perplexity is an indicator used to evaluate the quality of a language model and can be used to select the number of topics. Perplexity gradually decreases with increasing topics, so the lowest or inflection point in the perplexity curve corresponds to the optimal number of topics for a document. The perplexity calculation formula is shown in Formula (4).

$$Perlexity = \exp\left\{ -\frac{\sum_{i=1}^{M} \log p(wi)}{\sum_{i=1}^{M} Ni} \right\} \qquad (4)$$

The LDA topic model is used to extract topic features of the three-period microblog text. First, the perplexity is used to select the number of topics. It can be seen from the above formula that with the continuous increase of the number of topics, the perplexity value continues to decrease until it reaches minimum value. It can be seen from Fig. 2 that when the number of topics in the first period is N = 9, the perplexity is the smallest, and the model effect is optimal at this time. Similarly, the optimal number of topics in the second and third periods is N = 9 and N = 6.

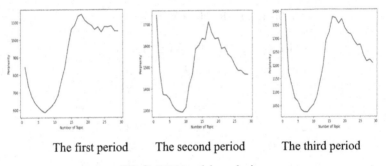

The first period   The second period   The third period

Fig. 2. LDA model perplexity

Table 2. Characteristic words of public topics during the lockdown period

Period	The theme	Key words
The first period	Topic0	Poetry recitation, special programs, day and night, dedication, heroes
	Topic1	More than 10000, sick, fight, live, patients
	Topic2	Resources, medical care, shortages, help, whatever it takes
	Topic3	Beds, patients, supplies, operations, shortages
	Topic4	Love, warmth, unity, sacrifice, warmth
	Topic5	Input, blockade, support, help, rescue team
	Topic6	Transplants, closures, bone marrow transplants, hospitals, heartbreaks

(continued)

Table 2. (*continued*)

Period	The theme	Key words
The second period	Topic0	After sealing the city, looking for live, sojourn, box, stranger
	Topic1	Master, send food, care, buy, dwelling
	Topic2	Power failure, lockdown, repair, epidemic, home
	Topic3	Sacrifice, life, kindness, crying, love
	Topic4	the city, fight the epidemic, struggle, overcome, retrograde
	Topic5	Wuhan, grounding, situation, persistence, prevention and control
	Topic6	Isolation, human history, virus, emergency, response
	Topic7	Forced, stranded, with relatives, trapped, home
	Topic8	Help, obligation, medicine delivery, voluntary, rebellious
The third period	Topic0	Flight, arrival, main city, checkpoint, sacrifice
	Topic1	Spread, helplessness, America, slander, developed countries
	Topic2	Door, photo, restart, control, zero hours
	Topic3	Nationwide, elite, Wuhan, containment, anti-epidemic
	Topic4	Before lockdown, 80%, COVID-19, undiagnosed, decisiveness
	Topic5	Report, first PI, details, Li Lanjuan disclosure
	Topic6	Detention, service, global, supreme inspection, retrograde
	Topic7	Backlight, corporations, donations, supplies, tens of millions

It can be seen from Table 2 that the first period was at the beginning of the "closed city", and a large number of infected people began to appear in the early period of the epidemic due to the influence of the virus. Due to insufficient understanding of the epidemic, the public panicked. From the topic analysis results, it can be seen that due to the large-scale outbreak of the epidemic, the shortage of medical resources is particularly prominent, the number of patients continues to increase, and there is a serious shortage of medical beds. Therefore, the most important issue in the early period is whether medical resources can meet the needs of patients.

Different from the first period, the second period of epidemic prevention and control has achieved results. With the continuous improvement of epidemic prevention and control measures, the most serious medical resource problems have been greatly solved. However, what followed was the shortage of material resources under the condition of citizens being closed and the resettlement of people who stayed in Han from other places.

After entering the final period of the lockdown, the public's problems with their own safety and needs for basic living materials have been basically resolved. Topic feature words show that the national epidemic has been effectively contained, and the restart of cities is gradually underway. However, the foreign epidemic is in a period of high outbreak, and the epidemic has spread to all parts of the world, which has aroused some public concerns about whether the domestic epidemic will break out again.

4.2 Analysis of Sentiment Evolution

Sentiment analysis, also known as propensity analysis and opinion mining, is a process of analyzing, processing, summarizing and reasoning on subjective texts with emotional color. Dividing emotions into binary (positive, negative) and ternary (positive, neutral, negative) for classification research, we can analyze the sentiment of the whole period of public opinion events. However, human emotions are complex, and their emotion classification cannot only stay at the positive and negative levels. It is necessary to subdivide emotions as much as possible to achieve the purpose of restoring the real emotions of netizens under public opinion events [27]. In the past, when scholars conducted sentiment analysis on text corpus, they were mostly limited to the emotional attributes of the text itself, and did not consider the emotional intensity of the emotional vocabulary itself. Therefore, this paper introduces the calculation of sentiment degree on the basis of sentiment subdivision, and analyzes the sentiment of netizens in more detail. The analysis and research were carried out according to the above three periods, and the experimental results are shown in Table 3 below.

Table 3. Distribution of public sentiment

Period of Emotion	The first period (2020.1.23–2020.2.21)	The second period (2020.2.22–2020.3.15)	The third period (2020.3.16–2020.4.8)
Anger	0.0135	0.0152	0.0406
Hate	0.2838	0.2523	0.1850
Fear	0.1529	0.0951	0.0831
Sad	0.0952	0.0937	0.0822
Surprised	0.0135	0.0076	0.0087
Happy	0.3422	0.4393	0.4676
Optimistic	0.0989	0.0968	0.1598

The first period (2020.1.23–2020.2.21): The multi-scale outbreak, large-scale spread and high contagiousness of the new crown epidemic, the Wuhan Municipal Government issued the "city closure" policy, and a large number of online comments spread panic information, which increased the popularity of topics related to the closure of the city and the pressure of public opinion have exacerbated the public's negative emotions. However, the ensuing Spring Festival fever reduced the attention on the topic of the epidemic. In addition, the national leaders delivered important speeches on the day of the Spring Festival, which greatly encouraged the public's confidence in fighting the epidemic. Public sentiment has recovered, and the proportion of positive sentiment has gradually increased.

The second period: With the relief of outstanding problems such as medical resources and living materials, the public's mood is basically in a stable state. From the results of the fine-grained analysis of sentiment, it can be seen that the proportion of the public's fear of the epidemic has dropped significantly due to various effective measures in my country to deal with the epidemic and relatively mature diagnosis and treatment plans. Moreover, the emergence of numerous "retroversaries" of the epidemic and the continuous support from all over the country made the proportion of the public's "good" sentiments rise rapidly, occupying a dominant position at this period.

The third period: On April 8, with the end of Wuhan's 76-day lockdown, the public's positive sentiment also peaked. The fine-grained analysis of emotions shows that in this period, anger occupies a certain proportion compared to the other two periods. The reason is that foreign media deliberately smear China's prevention and control work, and ignore China's role in the international community's epidemic work. The significant contribution made by the public has led to public outrage. However, as the unblocking date is approaching, various work has been restarted, and the proportion of the public's "beautiful" and "optimistic" emotions has increased, and positive emotions are still in the dominant position at this period.

5 Validation and Analysis of Prevention and Control Recommendations

Using the public opinion caused by the current regional epidemic—the Shanghai tour group event as the verification data set, the evolution trend of public sentiment in the dissemination of network public opinion is simulated through simulation experiments, and the impact of the recommendations in this paper on the distribution trend of public sentiment during the epidemic is verified. The SEI7R model is used to simulate the evolution trend of public sentiment in public opinion.

5.1 SEI7R Model

According to the characteristics of netizens holding different emotions in the communication of public opinion, the I state (infected people with public opinion information) in the traditional SEIR model is divided into seven types of emotionally infected persons: happy, good, angry, sad, all, evil, and shocked. At the same time, the following updates are made on the traditional SEIR transfer path: (1) The public opinion information infected persons in the model are divided into 7 types of infected persons according to their emotions, which are $I_1, I_2, I_3, I_4, I_5, I_6, I7$, and the public opinion information in the public opinion event is latent. The information infected person has a certain probability of being transformed into a state in the state of the information infected person I, and the increase path E is transformed into I_{1-7}. (2) Since the dissemination of information in public opinion is not the same as the dissemination of viruses, people susceptible to public opinion information may not necessarily experience a long incubation period after coming into contact with relevant information, and the susceptible person will have a certain probability to directly express it in the network. Own opinions on public opin-ion events, so the path S is transformed into I_{1-7} added to the SEIR model. (3) The difference between being a lurker in the communication of public opinion and the lurker after exposure to the virus is that the lurker after exposure to the virus cannot decide whether he is willing to be infected or directly transformed into an immune person, but the lurker in the communication of public opinion can Infected people or people immune to public opinion who can transform various emotions according to their own wishes, so increase the transformation path E to R. The model path transformation diagram is shown in Fig. 3.

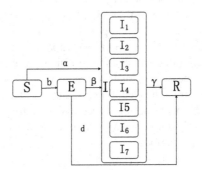

Fig. 3. SEI$_7$R model diagram

According to the division of the sentiment classification model in this paper and the SEI7R model, and combined with the characteristics of the Shanghai tour group event, the simulation experiment was carried out to construct the number of netizens = 1000. At the initial moment of the epidemic, 20 nodes were randomly selected as public opinion information infected persons. Combined with the characteristics of public opinion events, the initial parameters are set as: $E(0) = R(0) = 0$ $E(0) = R(0) = 0$, I1 (Disgust) = 8, I2 (Anger) = 1, I3 (Fear) = 2, I4 (Sad) = 2, I5 (Surprise) = 1, I6 (Beautiful) = 4, I7 (Optimism) = 2. S(susceptible) = 980, $\alpha 1 = 0.301$, $\alpha 2 = 0.019$, $\alpha 3 = 0.11$, $\alpha 4 = 0.107$, $\alpha 5 = 0.02$, $\alpha 6 = 0.188$, $\alpha 7 = 0.027$; $\beta 1 = 0.24$, $\beta 2 = 0.041$, $\beta 3 = 0.125$, $\beta 4 = 0.063$, $\beta 5 = 0.032$, $\beta 6 = 0.28$, $\beta 7 = 0.102$; $\gamma 1 = 0.2$, $\gamma 2 = 0.293$, $\gamma 3 = 0.1353$, $\gamma 4 = 0.1066$, $\gamma 5 = 0.1463$, $\gamma 6 = 0.1258$, $\gamma 7 = 0.0997$, d = 0.117, b = 0.228. the model simulation diagram is shown in Fig. 4.

Throughout the entire process of public sentiment and topic evolution during the lockdown period, public sentiment, public opinion topics, and prevention and control measures developed cyclically and synchronously, and there was an interaction relation-ship. On the one hand, the emergence and spread of public opinion topics will affect the trend of public sentiment, and the trend of public sentiment will prompt the govern-ment to disclose information in a timely manner and optimize prevention and control measures; on the other hand, the optimization of prevention and control measures and the disclosure of prevention and control information will be curbed in time. The further development of rumors can prevent the spread of negative public opinion events, thereby eliminating the panic in the public and the growth of negative emotions. Therefore, this paper puts forward three suggestions for the prevention and control of public opinion for

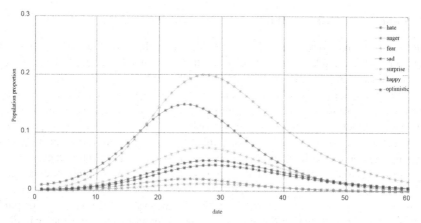

Fig. 4. Initial simulation diagram

the incubation period, mature period and outbreak period of public opinion development and carries out simulation verification respectively.

5.2 Strengthen Network Information Supervision

Online public opinion on public health emergencies has comprehensive characteristics such as super-large-scale information, rapid information release and dissemination, large participating groups, and strong real-time interaction. Therefore, strengthening network information supervision can play a certain guiding role in the trend of public opinion events, and can promote the transformation of public opinion information infected I and lurker E to public opinion information immune R. The simulation experiment is carried out on this, and the conversion probability of e $e(I \rightarrow R)$, d $(E \rightarrow R)$ is increased by 20%, and the simulation diagram is shown in Fig. 5.

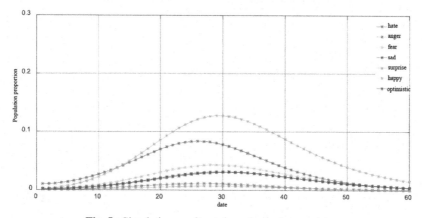

Fig. 5. Simulation results under enhanced supervision

5.3 Improve the Guiding Ability

When public opinion develops to a certain extent, the public opinion event itself and the characters in the event have been formed, and the supervision mechanism can no longer play a role in it. At this period, the official media and government platforms should intervene and guide, actively report and actively guide public sentiment by virtue of their own speed of dissemination and authoritative characteristics, eliminate the panic and anxiety generated by the public under public opinion, and make it as positive as possible in public opinion events. Timely information disclosure can control the trend of public opinion events in advance, and early management and control can control the increase rate of public opinion infected people in time. In the simulation experiment, the control measures were released on the 21st day, so the conversion rate of α_1, α_3, α_4, α_4 and γ_1, γ_2, γ_3, γ_4 was reduced by 20% while accelerating the conversion of public opinion infected people to immunized people. The simulation diagram is shown in Fig. 6.

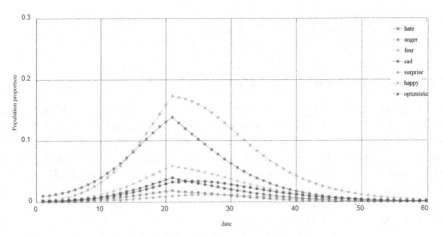

Fig. 6. Simulation results under advanced control

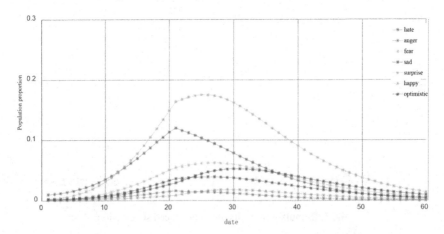

Fig. 7. Simulation results when reporting positive news

5.4 Report Socially Positive Hot News

The themes of online public opinion are extremely extensive and are characterized by diversity and suddenness. It is difficult to control public opinion events in a timely manner only by relying on the network influence of official media. In-depth cooperation with other mainstream media and network leaders is necessary. Positive social hot news can increase the proportion of people with positive emotions, and promote the conversion probability of susceptible and latent groups to positive emotions. In the simulation experiment, the conversion rate of $\beta4$, β_5, β_6 and γ_4, γ_5, γ_6 20% is increased. The simulation experiment is shown in the following Fig. 7.

6 Conclusion

This paper takes the comments of netizens on Weibo during the new crown epidemic as the research object, and analyzes the interaction between the characteristics of public opinion topics and the fluctuations of netizens' emotions during the period of the epidemic. Firstly, the related comment data of Weibo is processed, and the improved BERT-BiLSTM model is used to analyze the sentiment of the preprocessed Weibo corpus, and the sentiment dictionary is introduced to divide the sentiment in a fine-grained manner. Emotional state. Secondly, on the basis of sentiment analysis, LDA model is used to extract themes of each period of the epidemic, trying to find key events and nodes related to public sentiment fluctuations during the epidemic. Finally, it analyzes the government response level, and puts forward corresponding countermeasures and suggestions for each period of the epidemic. By using the improved SEI7R model to simulate and verify the data of Shanghai tour group public opinion events, it analyzes the incubation period, mature period and outbreak period of public opinion respectively. Means. The experimental results prove the effectiveness of the recommendations proposed in the study on the prevention and control of public opinion, and provide reference and theoretical support for the governance of public opinion on the Internet of public health emergencies that may occur in the future.

References

1. Hxa, C., Wa, A., Jl, B., et al.: Outlier knowledge management for extreme public health events: understanding public opinions about COVID-19 based on microblog data. Socio-Econ. Plan. Sci. **80**, 100941 (2020)
2. Zhang, D., Wei, J.B.: Epidemic information data analysis and discourse guidance strategy of mainstream media driven by emotion. Libr. Inf. Serv. **65**(14), 101–108 (2021)
3. Shigemura, J., Ursano, R.J., Morganstein, J.C., et al.: Public responses to the novel 2019 coronavirus (2019-nCoV) in Japan: mental healthconsequences and target populations. Psychiatry Clin. Ne-urosci. **74**(4), 281–282 (2020)
4. Muni, Z., Yong, L., Xu, T., et al.: Online public opini-on evolution simulation of COVID-19 based on Bert-LDA model. J. Syst. Simul. **33**(01), 24–36 (2021)
5. Zhongbao, L., Quan, Q., Wenjuan, Z.: Analysis of the impact of COVID-19 on Netizens' emotions in the micro-blog environment. J. Inf. **40**(02), 138–145 (2021)
6. Shujin, C., Wenyu, Y.: Mining and evolution analysis of microblog public opinion to-pics in public health emergencies. J. Inf. Resour. Manage. **10**(06), 28–37 (2020)

7. Dan, W., Haitao, Z., Yashu, L., Liang, R.: Emotional tendency analysis and thought leading research on key nodes of microblog public opinion. Libr. Inf. Serv. **63**(04),15–22 (2019)

8. Jian-xia, C., Jun-yi, L.: Research on the temporal and spatial differentiation of COV-ID-19 epidemic and public anxiety: based on microblog data. Hum. Geogr. **36**(03), 47–57+166 (2021)

9. Xi-wei, W., Yue-qi, L., Cheng-cheng, Q., Huan, H.: Research on reversal model and simulation of network rumor propagation under public health emergencies. Librar. Inf. Serv. **65**(19), 4–15 (2021)

10. Hu, M., Liu, B.: Mining and summarizing customer reviews. In: Proceedings of SIGKDD International Conference on Knowledge Discovery and Data Mining, pp. 168–177 (2004)

11. Yang, J., Leskovec, J.: Modeling information diffusion in implicit networks. In: IEEE International Conference on Data Mining, IEEE (2011)

12. Strapparava, C., Valitutti, A.: Wordnet affect: an affective extension of wordnet. Lrec. **4**(1083–1086), 40 (2004)

13. Neviarouskaya, A., Prendinger, H., Ishizuka, M.: SentiFul: a lexicon for sentiment analysis. IEEE Trans. Affect. Comput. **2**(1), 22–36 (2011)

14. Lin-hong, X., Hong-fei, L., et al.: The construction of affective vocabulary ontology. J. China Soc. Sci. Technol. **27**(2), 180–185 (2008)

15. Blair-Goldensohn, S., Hannan, K., et al.: Building a sentiment summarizer for local service reviews. In: WWW Workshop on Nlp Challenges in the Information Explosion Era (2008)

16. Whisner, C.M., Wang, H., Felix, S., et al.: Mining the Twitter-sphere for consumer attitudes towards dairy. FASEB J. **30**, 897.2–897.2 (2016)

17. Delan, X., Huming, C., Shengli, T.: A study on sentence commendatory or derogatory orientation based on HowNet. Comput. Eng. Appl. **22**, 143–145 (2008)

18. Feng-Ying, H.E.: Orientation analysis for Chinese blog text based on semantic comprehension. J. Comput. Appl. **31**(08), 2130 (2011)

19. Pang, B., Lee, L., Vaithyanathan, S.: Thumbs up? Sentiment classification using machine learning techniques. In: Proceedings of the ACL-02 Conference on Empirical Methods in Natural Language Processing, Philadelphia. USA, pp. 79–86 (2002)

20. Chun-guang, B., Shuai, Y., Ke, H., Hai, G., Jin-long, W.: Analysis of sentiment Orientation of multiple characteristics based on ginseng purchase comments. Northeast Agric. Sci. **45**(03), 92–96 (2020)

21. Songtao, S., Yanxiang, H.: Multi-label sentiment classification based on CNN feature space. Adv. Eng. Sci. **49**(03), 162–169 (2017)

22. Hu, C.J., Liang, N.: Thematic emotion analysis based on LSTM of Deep attention. Appl. Res. Comput. **36**(04), 1075–1079 (2019)

23. Peng, W., Ting, L., Chong, T., Si, S.: Research on emotion classification of financial microblog based on OCC model and LSTM model. J. China Soc. Sci. Technol. **39**(01), 81–89 (2020)

Failure Simulation Study of Double-Layer Exhaust Manifold for Turbocharged Engine Under LCF & HCF

Dongzhe Xuan[1(✉)], Yanji Piao[2], Dechao Wang[1], and Zhijun Quan[3]

[1] College of Engineering, Yanbian University, Jilin 133002, China
xuandongzhe@ybu.edu.cn
[2] Information Management and Information System, Yanbian University, Jilin 133002, China
[3] College of Information and Electrical Engineering, China Agricultural University,
Beijing 100083, China

Abstract. With the high requirements of turbocharged engine fuel efficiency and high exhaust gas temperature and continuous improvement of emission standards, the structure of automobile exhaust manifold needs to be continuously optimized on the premise of ensuring durability and reliability. Taking the exhaust manifold of a four cylinder turbocharged engine as the research object, the low cycle fatigue and high cycle fatigue at high temperature were studied as follows. Firstly, Hyper-mesh is used for modeling, STARCCM + is used for CFD calculation, and ABAQUS is used for thermal stress calculation and thermal mode calculation under standard conditions. The failure point of the system is preliminarily judged according to the FEM calculation results, and then compared with the durability test results to verify the reliability of the FEM calculation results. Finally, the optimal design scheme of exhaust manifold is proposed and verified by simulation calculation. The results show that the DPEEQ value of the outer ring is 1.06%. The elastic strain energy under thermal mode calculation is 49.4J, which is consistent with the cracking point in the durability bench test results. By optimizing the outer ring design, the elastic strain energy is reduced by 20.9%. Our research provide theoretical support and reference for the analysis method of exhaust manifold failure under low cycle fatigue and high cycle fatigue.

Keywords: Exhaust manifold · LCF · HCF · Thermal stress · Thermal mode

1 Introduction

In order to meet people's requirements for fuel efficiency of turbocharged engine and domestic environmental protection standards for exhaust gas, the innovative design and development of exhaust manifold as one of the key components of automobile has become the focus of attention in the automotive field. Therefore, the automobile industry at home and abroad has introduced new materials, new processes, innovative design and manufacturing technology, which is expected to achieve the goal of reducing engine size while ensuring power, also reducing exhaust pollution and fuel consumption. However,

© The Author(s), under exclusive license to Springer Nature Singapore Pte Ltd. 2023
S. Yang and S. Islam (Eds.): APWeb-WAIM 2022 Workshops, CCIS 1784, pp. 125–137, 2023.
https://doi.org/10.1007/978-981-99-1354-1_12

the premise of the above objectives is that how to ensure the durability and reliability of the exhaust manifold under the dual conditions of high cycle fatigue and low cycle fatigue is the difficulty in its optimization process.

In the process of automobile development, the design space of exhaust manifold is limited due to the assembly of engine and other components. In order to ensure the dynamic performance of fluid in exhaust manifold and the durability of structure, automobile manufacturers will usually be designed the manifold into complex geometry. Therefore, the complex exhaust manifold structure will directly reduce the fatigue life and lead to the products failure under the impact of high temperature and various loads.

The reduction of fatigue life can be considered from three ways, material characteristics and structural design, low cycle fatigue(LCF), high cycle fatigue(HCF). At the exhaust gas temperature above 800~1000 °C, the physical and mechanical properties and fatigue characteristics of cast iron, cast steel, ferrite-austenite stainless steel is different. Compare with other materials, austenitic materials show better fatigue resistance at high temperature under the heat transfer test, stress-strain test and low cycle fatigue test [1, 2]. In general, the manifold uses cylindrical stainless steel, but the each point thickness of tube will be different according to the complex structure and processing technology such as rolling or stamping, which will shorten the fatigue life of the manifold [3].

Under the long-term repeated impact of high-temperature alternating load, the exhaust manifold has thermal stress and plastic strain lead to cracking and damage of the manifold [4, 5]. Therefore, many scholars use the fluid structure coupling simulation, take STARCCM + and ABAQUS software as the calculation platform, through the finite element analysis to calculate temperature field and stress field distribution under the alternating cold and hot working conditions, and make the comparison between the finite element calculation and the durability bench test results, verify the feasibility of using the simulation calculation to analyze the fatigue failure of exhaust manifold [6]. In the verification process, the low cycle fatigue analysis theory and Coffin-Manson fatigue life calculation method are used as the theoretical basis to improve the accuracy of manifold life prediction. Finally, it is proposed that under the condition of low cycle fatigue, the failure prediction can be judged according to the delta equivalent plastic strain (DPEEQ) in the simulation results [7–9].

The modal frequency values of the manifold are different at room temperature and high temperature. The high cycle fatigue phenomenon caused by vibration load at high temperature condition that is also another main factor affecting the damage or cracking of the manifold [10]. The increase of exhaust gas temperature causes nonlinear changes in mechanical properties of materials, such as elastic module, heat transfer coefficient and expansion coefficient. Therefore, the influence of temperature on the vibration characteristics of the manifold must be considered. Multi-objective optimization can be carried out through topology optimization and strengthening structural stiffness to improve the vibration fatigue life of the manifold and improve the phenomenon of low cycle fatigue [11, 12]. However, the crack of the exhaust manifold of turbocharged engine is caused by superposition of thermal and vibration stress. Therefore, the influence of engine vibration on the durability of the exhaust manifold should also be considered at high temperature [13]. The fatigue life can be analyzed by the stress field under the thermal and vibration load in the calibration condition and the failure point of the manifold can be judged from the thermal and vibration calculation results, which provides a theoretical basis and reference for the life prediction and structural optimization of the exhaust manifold [14, 15].

The above results show that some scholars have analyzed the shortening of exhaust manifold fatigue life from the material viewpoint, others have proposed the influence of plastic deformation on low cycle fatigue life, and some scholars have studied the structural failure of high cycle fatigue caused by vibration characteristics. All scholars have studied the durability of exhaust system from one or two sides, but the reduction of fatigue life is from the complex load and multiple influences. Therefore, this paper takes the double-layer structure exhaust manifold of a four cylinder turbocharged engine as the research object to verify the influence of thermal load and vibration load on failure through calculation of nonlinear finite element method.

2   Exhaust Manifold Failure Analysis Method and Process

As shown in Fig. 1, two models are established. First one is the fluid mesh model (CFD model) and the other is the structural mesh model (FEM model). According to different calculation methods give different boundary conditions and material properties. The temperature distribution and heat transfer coefficient of the internal flow field in the manifold are calculated by CFD software to analyze the accuracy of the temperature field and can be used for next structural heat transfer calculation. The analysis of low cycle fatigue and high cycle fatigue are carried out through thermal stress and thermal mode calculation. Map the temperature from the CFD results for heat transfer calculation, and then add the calculated temperature results to the thermal stress model to calculate the stress distribution and equivalent plastic strain of each component. Similarly, the calculation of thermal mode should also map the temperature in the finite element model and calculate the heat transfer, and then calculate the natural frequency and elastic strain energy to judge the failure point. The calculation results need to be compared with the electron microscope analysis results of the cracked manifold in the durability bench test, and verify the similarity of the both results. Finally, optimize the structure of the cracked parts of the manifold and iterating the finite element calculation.

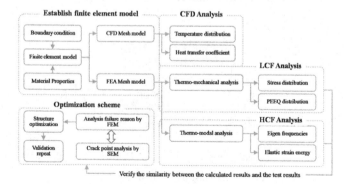

Fig. 1. Exhaust manifold failure analysis flow chart

3   Establish Finite Element Model

CFD Mesh Model

At present, the CFD calculation of exhaust manifold generally use the hydrodynamics calculation software STARCCM +, which imports the 3D model of exhaust manifold into Hyper-mesh with good interchangeability with STARCCM + to generate the mesh of surface, and then imports that into STARCCM + for surface remesher, prism layer mesher, polyhedral mesher etc. Generate the hexahedral based polyhedral mesh with a 3mm base size and a one layer 1mm prism layer [16]. The turbulence model selects the Realizable $k - \varepsilon$ Two-Layer model of RANS algorithm, which is widely used in the industry when calculating the fluid dynamics of exhaust manifold and the wall function uses the Two—Layer All y + Wall Treatment no-slip boundary condition [17, 18]. Set the mass flow and temperature at the inlet and pressure at the outlet, set the convergence criterion as 4000 steps at the maximum iteration step, 0.01 asymptotic limit of flow and 0.05 asymptotic limit of pressure. The input boundary conditions are shown in Table 1.

Table 1. Boundary conditions for CFD calculation of exhaust manifold

Mass flow(kg/h)	Temperature(°C)	Pressure(mbar)
360	950	500

FEM Mesh Model

The finite element model of exhaust manifold is different according to the type of calculation method such as heat transfer calculation, thermal stress calculation and thermal mode calculation. Firstly, 3D software is used for surface modeling, and then Hyper-mesh is used to establish the mesh model. The mesh quality and the material properties have a direct influence on the calculation results. The established thermal stress calculation model and thermal mode calculation model are shown in Fig. 2.

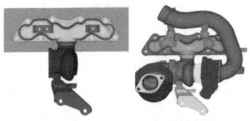

(a) Thermal stress calculation model   (b) Thermal mode calculation model

Fig. 2. Exhaust manifold mesh model

- Mesh quality: The mesh size is basically 2mm to ensure the calculation accuracy. Tubes and housing made by shell element, and solid element used for inlet flange, bracket, welding part and outer ring.
- Material properties: For the main material SUS309S, the specific heat is 500J/kg°C, the Poisson ratio is 0.28 and the density is 7900kg/m^3. As shown in Fig. 3, the yield strength and tensile strength are changed at different temperatures. Other parameter such as heat transfer coefficient, young's modulus and expansion coefficient also shows different characteristics at different temperatures as shown in Table 2.

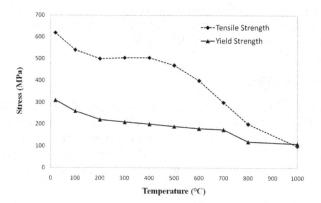

Fig. 3. Material stress curves at different temperatures

Table 2. Material parameters at different temperatures

Temperature (°C)	Heat transfer coefficient(W/m°C)	Young's modulus (GPa)	Expansion coefficient(10^{-6}/°C)
20	15.0	200	15.8
100	15.5	196	16.5
200	17.5	185	17.0
400	20.0	170	18.0
1000	28.5	120	19.5

4 Low Cycle Fatigue Analysis

4.1 Theory

The low cycle fatigue life the numbers of thermal load cycles for plastic deformation and leads to manifold fracture under the repeated cycle of stress and strain. With the increase of the repeated thermal load cycle numbers, the stress amplitude and plastic strain amplitude become smaller and the area of hysteresis loop becomes smaller and finally the manifold will crack and expand until break [6]. This number of cycles is the material life, which can be estimated by Manson-coffin formula [3], as follows:

$$\frac{\Delta\varepsilon_t}{2} = \frac{\sigma_f^{'}}{E}\left(2N_f\right)^b + \varepsilon_f^{'}\left(2N_f\right)^c \tag{1}$$

where $\Delta\varepsilon_t$ is total strain amplitude; $\sigma_f^{'}$ is fatigue strength coefficient; N_f is number of failure cycles; $\varepsilon_f^{'}$ is fatigue ductility coefficient; E is elastic modulus; b is fatigue strength exponent; c is fatigue ductility exponent. The fatigue strength exponent and fatigue ductility exponent in the formula can be obtained through the accumulation of test data. Generally, research institutes or laboratories will build relevant database for reference when developing similar models in after.

4.2 Boundary Conditions

The temperature distribution and heat transfer coefficient of the internal flow field are calculated by CFD, and the temperature field of the structure surface is obtained after mapping the FEM model. The fixed part of the turbocharger bracket need constrain six DOF. The intake flange part is connected on the engine block model and applies the bolt preload, and the engine block need to be restrained in three directions, as shown in Fig. 4.

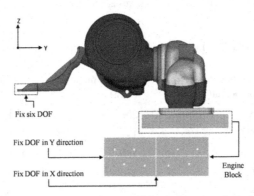

Fig. 4. Boundary condition diagram of thermal stress calculation model

The simulation calculation steps should be consistent with the engine bench test steps. Therefore, the working conditions used in this paper are shown in Fig. 5. Firstly, pre tighten the bolt, increase the temperature, cooling down the engine and then increase the temperature again to calculate the delta equivalent plastic strain (DPEEQ) during the second heat up and the first cooling down.

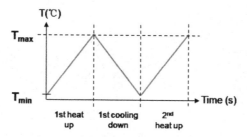

Fig. 5. Operating diagram of thermal stress calculation

4.3   Results

The heat transfer calculation results of double-layer exhaust manifold are shown in Fig. 6. The highest temperature is 925.9 °C occurs at the joint of inner tube and outer ring. In the figure, the engine block and intake pipe are contact with the surrounding environment, so their temperature should set to be little higher than the surrounding environment. There will also be heat conduction, convection and radiation between the shell and the internal tubes, so there also has significant high temperature on the shell. The outer ring combined with the turbocharger at the joint of the inner tubes. Subject to four cylinder engine, the outer ring will be impacted by high-temperature exhaust gas in all working cycles, so it shows the highest temperature and may also be a potential failure point.

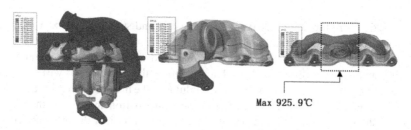

Max 925. 9℃

Fig. 6. Temperature distributions of manifold

The physical meaning of equivalent plastic strain (PEEQ) is an amount the cumulative value of plastic strain. From the mechanical fatigue characteristics of materials, under the impact of high temperature and low temperature alternating load, the compressive stress and tensile stress will be repeated transformation and plastic strain will occur when the stress over the yield strength limit. According to the stress-strain hysteresis loop, the original state cannot be restored after plastic strain has occurred. Eventually, the manifold will be cracked and the crack will extend after the impact cycle and continuous accumulation of plastic strain.

The thermal stress calculation results are shown in Fig. 7. The bolt hole of the turbocharger bracket fixed with six degrees of freedom, so that shows a high DPEEQ value 2.0%. According to experience the high plastic strain value caused by hard constraints, that is can be ignored. The DPEEQ values of the inner tube and outer ring are 0.53%

2.0 %   0.53 %   1.06 %

Fig. 7. DPEEQ distributions

and 1.06%, which are close to the weld part and the result over the industry allowable value 1%. According to the thermal stress calculation results, the fracture probability of inner tube and outer ring will be very high.

5 High Cycle Fatigue Analysis

5.1 Theory

According to the mechanical vibration theory, the structure is usually regarded as an elastic coupling rigid body composed of mass point, rigid body and damping. The vibration differential equation [22] is:

$$M\ddot{x} + C\dot{x} + Kx = F \tag{2}$$

where M, C, K are the mass matrix, damping matrix and stiffness matrix; x, F are the displacement vector and excitation load vector.

The Eigen frequency values of exhaust manifold under normal temperature and operating state are very different. At high temperature, the thermal load will change the stiffness of the structure, because the elastic modulus of the material will decrease at high temperature, which will directly affect the stiffness distribution and change the stiffness matrix, as shown in Table 2. Therefore, it is necessary to consider the influence of high temperature environment on vibration fatigue characteristics in modal calculation.

5.2 Results

Within the economic speed range, if the Eigen frequency of the engine is close to the Eigen frequency of the exhaust manifold, the resonance will lead to the crack of the exhaust manifold. Therefore, it is necessary to avoid the Eigen frequency of exhaust manifold in the frequency range 27–200 Hz (800–6000 rpm/min). The modal calculation results are shown in Fig. 8. The first and second mode frequencies are within the frequency range of economic speed and there has potential risk of resonance. In order to further confirm the potential failure point, the distribution of elastic strain energy under the first two Eigen frequencies was analyzed. The maximum value occurs in the outer ring which is 49.4J and 18.0J at first and second modes.

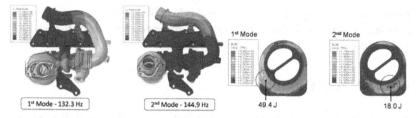

Fig. 8. Eigen frequencies and elastic strain energy of first and second mode

6 Failure Point Examination

In order to verify the reliability and durability of the exhaust manifold, the durability bench test was carried out according to the national standard and the OEM performance test method for the durability of the exhaust manifold. The working condition is based on the test standard provided by an OEM. After more than 700 cycles, the engine performance was decreased and the exhaust manifold has cracked. Through the leakage test, there found the leakage point occurs in the outer ring, and the crack position is consistent with the position where the maximum elastic strain energy occurs in the vibration calculation. It can be preliminarily judged that the crack is caused by vibration. From the shape of the crack, we also can judge whether the cause is high cycle fatigue or low cycle fatigue. Through the electron microscope analysis of the failure part, the outer ring crack is mainly comes from thermal cycle load and local stress, and the crack is accelerated by high temperature and vibration (Fig. 9).

Fig. 9. Leakage test and electron microscope analysis of outer ring crack point

7 Structural Optimization Scheme

In engineering design, optimization design is to choose simple, effective and direct way among many schemes as much as possible. The most effective method of structural optimization is to combine the finite element calculation method and test method, and refer the test results to modify the finite element model for verification. According to the comparison of the two results, two structural optimization schemes are obtained in this paper. Firstly from the thermal stress side, the plastic deformation can be reduced by changing the structure of the inner tube and the thickness of the outer ring; Secondly from the thermal mode side, the structural stiffness can be improved by changing the design of the bracket and outer ring. Through the comparison of the common points of the two schemes, the simple and direct method is to optimize the design of the outer ring which is increase the thickness of the bottom and connecting part of the outer ring to improve the structural stiffness, as shown in Fig. 10.

8   Verification of Optimization Scheme

In order to reduce the research cycle and cost, the optimized exhaust manifold is only verified by finite element calculation. The thermal stress calculation results are shown in Fig. 11. The result of the turbocharger bracket does not change much, but still shows a high DPEEQ value 1.95%, and the DPEEQ value of the inner tube part is 0.54% which is similar to the before optimization result. However, the DPEEQ value of the outer ring is reduced by 0.39% and the result is within the allowable value range 1% which can be judged as no potential risk of failure.

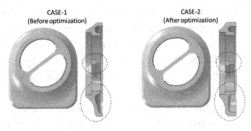

Fig. 10. Optimization design of outer ring structure

The modal calculation results of the optimized outer ring are shown in Table 3. The Eigen frequency value is basically same compared with before optimization. However, the elastic strain energy at the first mode is reduced by 20.9% from 49.4J to 39.1J compare with before optimization, and the elastic strain energy at the second mode is reduced by 25% from 18.0J to 13.5J as shown in Fig. 12 and Table 3. The vibration characteristics of the optimized structure are greatly improved.

Fig. 11. DPEEQ distribution of optimized design

Through the verification of finite element calculation, it can be judged that the optimization design of outer ring has a significant effect to improve the durability of exhaust manifold, and this scheme is feasible.

Table 3. Comparison of eigen frequency values and elastic energy values

	Eigen Frequencies(Hz)		Elastic strain energy(J)	
	CASE-1	CASE-2	CASE-1	CASE-2
1st Mode	132.3	132.5	49.4	39.1
2nd Mode	144.9	144.7	18.0	13.5

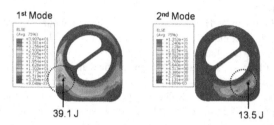

Fig. 12. Elastic strain energy distribution of optimized design

9   Conclusions

In studying the durability or fatigue life of exhaust manifold, we can't only consider a single aspect. We need to comprehensively analyze and judge coupling effect of the various causes and between various factors.

- The results of thermal fatigue calculation show that the local plastic strain occurs under the alternating load of high and low temperature, and finally accelerate the manifold breaks. The DPEEQ value of the outer ring is 1.06% which is over the allowable value and has potential risk. The reliability of the test product is directly related to the welding conditions. Therefore, the processing quality should be strictly managed before the durability bench test.
- The modal calculation results show that the Eigen frequencies of the first two modes are 132.3 Hz and 144.9 Hz, which are within the frequency range of economic speed, and there has potential risk of resonance. The elastic strain energy at the first two modes is 49.4J and 18.0J, which is reduced by 20.9% and 25% through design optimization. The optimized design improves the durability of the manifold and proves the feasibility of the optimization scheme.
- The dual loads of thermal stress and vibration at high temperature need to be considered when analyzing the failure of exhaust manifold by finite element method. The test results show that the failure reason is caused by comprehensive fatigue.

Therefore, it is proved that the cross and comprehensive analysis of finite element calculation results and test results and then propose the optimization scheme is a typical method to study the failure of exhaust manifold at recent stage. It is also the most effective and direct research method, and also provides guidance for the durability research of exhaust manifold.

Acknowledgements. This work was supported by Application Foundation Program of the Science and Technology Development Plan of Yanbian University of China (NO. 602020024).

References

1. Ekström, M., Jonsson, S.: High-temperature mechanical and fatigue properties of cast alloys intended for use in exhaust manifolds. Mater. Sci. Eng. A. (2014). https://doi.org/10.1016/j.msea.2014.08.014
2. Benoit, A., Maitournam, M.H., Rémy, L., et al.: Cyclic behaviour of structures under thermo-mechanical loadings: application to exhaust manifolds. Int. J. Fatigue **38**, 65–74 (2012). https://doi.org/10.1016/j.ijfatigue.2011.11.012
3. Gabellone, D., Plano, S.: LCF and TMF on ferritic stainless steel for exhaust application. Procedia Eng. **74**, 253–260 (2014). https://doi.org/10.1016/j.proeng.2014.06.258
4. Quan, Y.U.A.N., Xiaoyan, W.A.N.G., Hongbo, J.I., et al.: Simulation study of thermo-mechanical fatigue failure of diesel exhaust manifold. Chin. Intern. Combust. Eng. Eng. **41**(03), 87–92 (2020)
5. Weijin, G.A.O., Ting, C.H.E.N., Yingtao, X.U.: Study on crack failure of the exhaust manifold based on FEA and CFD. Automobile Parts. **12**, 30–33 (2020)
6. Kaimin, L., Jing, Y., Siyuan, Z., et al.: Low cycle fatigue study of turbocharged direct injection engine exhaust manifold. Autom. Eng. **38**(03), 373–379+384 (2016)
7. Ze-hao, H., Jing-rong, H., Xian-long, T.: On fluid-solid coupling thermal simulation analysis of exhaust manifold. J. Southwest China Normal Univ. (Nat. Sci. Edn). **45**(06), 107–112 (2020)
8. Hanyu, Z., Shanhu, Y., Benchao, W., et al.: Thermal fatigue life prediction of an exhaust manifold made of high nickel cast iron. Autom. Eng. **39**(08), 943–950+934 (2017)
9. Yuanying, L.: Thermal fatigue simulation analysis for the double-layer water cooled exhaust manifold of engine. North University of China (2015)
10. Rajadurai, S., Guru Prasad, M., Kavin, R., Sundaravadivelu, M.: Modal analysis for exhaust manifold in hot condition, Is there a need. SAE Technical Paper 2014-28-0036 (2014). https://doi.org/10.4271/2014-28-0036
11. Junhong, Z., Yusheng, Z., Jian, W., et al.: Vibration characteristics and fatigue life of a gasoline engine's exhaust manifold under high temperature environment. J. Vib. Shock. **36**(13), 33–40 (2017)
12. Junhong, Z., Yusheng, Z., Jian, W., et al.: Multi-objective optimization design for exhaust manifold considering thermo-mechanical coupling. J. Zhejiang Univ. (Eng. Sci.). **51**(06), 1153–1162 (2017)
13. Sissaa, S., Giacopinia, M., Rosia, R.: Low-cycle thermal fatigue and high-cycle vibration fatigue life estimation of a diesel engine exhaust manifold. Procedia Eng. **74**, 105–112 (2014). https://doi.org/10.1016/j.proeng.2014.06.233
14. Yuliang, X., Wei, L., Zhen, W., et al.: Study on the effect of engine vibration on low cycle fatigue life of exhaust manifold. Chin. J. Eng. Des. **25**(03), 330–337 (2018)
15. Xiangrun, M.: Analysis and solution of cracking of exhaust manifold of turbocharged engine. Intern. Combustion Eng. Parts. **09**, 91–93 (2017)
16. Zhijie, G., Tingting, Z., Qixian, G., et al.: Optimization design of double-layer exhaust manifold structure for turbocharged gasoline engine. Intern. Combust. Eng. 03, 24–28+37 (2016)
17. Nan, L., Chenhong, Z., Yingming, H., et al.: Analysis of intake manifold uniformity and selection of preferred structural parameters for gasoline engine. Veh. Eng. 01, 28–33 (2021)

18. Yingming, H., Chuhua, H., Zhuo, L.: Flow field investigation and optimization of intake manifold runner by CFD Method. Autom. Parts **01**, 42–45+58 (2014)

19. Jun, L., Lei, W., Lu, X.: Based on star-ccm+ steady CFD analysis of the intake manifold. J. Xiangtan Univ. (Nat. Sci. Edn.) **37**(02), 97–101 (2015)

20. Shuaishuai, Z., Yongxiang, C., Yening, J., et al.: Design and assessment of accelerated life testing based on modified Coffin-Manson model. Struct. Environ. Eng. **40**(04), 52–58 (2013)

21. Xiaoyu, L.: Study on Design Theory of an Automobile Exhaust Manifold. Wuhan University of Technology (2013)

22. Dongzhe, X., Yanji, P., Dechao, W.: Comprehensive analysis of modal and static calculation for automotive exhaust system. J. Wuhan Univ. Technol. **43**(02), 23–27 (2021)

Lung Electrical Impedance Tomography Based on Improved Generative Adversarial Network

Xiuyan Li📧, Ruzhi Zhang(✉)📧, Qi Wang📧, Xiaojie Duan📧,
and Jianming Wang📧

Tiangong University, Tianjin 300387, China
rzhizhang@163.com

Abstract. The image reconstruction of electrical impedance tomography (EIT) is highly ill-posed and nonlinear, the reconstructed images tend to have artifacts due to noise in the measurement system. Although deep neural networks have demonstrated great potential to remove artifacts from initial conductivity images, the interpretability and generalization ability of the network is difficult to guarantee. A deep learning structure, namely, a conditional generative adversarial network with an attention mechanism and residual connection (CGAN-AMR), is proposed for EIT image reconstruction. The attention mechanism is utilized in the generator to learn channel dependencies, and residual connection is employed by the discriminator to improve training efficiency. As a result, the accuracy and interpretability of CGAN-AMR are improved compared with CNN and CGAN methods in the EIT imaging task. The imaging results indicate that CGAN-AMR structure can effectively improve the clarity of the reconstructed lung images, the location and boundary of lung lesions are restored accurately.

Keywords: Electrical impedance tomography · Image reconstruction · Generative adversarial network · Lung disease

1 Introduction

Lung disease is one of the malignant diseases with the fastest increase in morbidity and mortality and the greatest threat to human health and life. In the past fifty years, many countries had reported significant increases in the morbidity and mortality of lung cancer. Therefore, early prevention is extremely important for the treatment of lung diseases. Different from CT and MRI which are expensive, time-consuming, and radiation hazardous, electrical impedance tomography (EIT) has broad development prospects as a non-invasive medical diagnosis method in the diagnosis of lung diseases. Electrical impedance

This work is supported by National Natural Science Foundation of China (61872269, 61903273, 62072335, 62071328) and Tianjin Natural Science Foundation (No. 19JCY-BJC16200).

S. Yang and S. Islam (Eds.): APWeb-WAIM 2022 Workshops, CCIS 1784, pp. 138–150, 2023.
https://doi.org/10.1007/978-981-99-1354-1_13

tomography technology is an electrical tomography (ET) technology that reconstructs the conductivity distribution of the sensing domain through the measured boundary voltages. The image reconstruction of EIT is a serious ill-posed nonlinear inverse problem [1]. EIT reconstructed images usually suffer from distortion and low resolution, which restricts the development of EIT technology. In order to improve the image quality, the linear back-projection (LBP) algorithm [2], Landweber iteration algorithm [3], Newton-Raphson algorithm [4] and Tikhonov algorithm [5] were proposed. Although these reconstructed algorithms have greatly improved the image quality, the blur and lack of detailed features of the reconstructed images are still the key factors restricting their application. The existing reconstruction methods are based on linearization of the EIT inverse problem, which is difficult to overcome its inherent shortcomings of strong non-linearity and serious ill-posed.

In recent years, deep learning has flourished in the field of biomedical image processing [6]. In [7], CNN provided a reference model for iterative deterministic optimization so that the reconstructed image satisfies both data consistency and prior knowledge. In [8], residual CNNs were trained to learn the aliasing artifacts brought by down-sampled MR brain data. In [9], GAN with pixel-wise loss, frequency domain loss, and perceptual loss was proposed for high-resolution MRI reconstruction.

The use of deep learning to improve the quality of image reconstruction has become an important direction of EIT research. An autoencoder neural network has been employed in electrical capacitance tomography (ECT) [10]. Convolutional neural networks (CNNs) have been used in EIT [11] and ERT [12]. A real-time capable reconstruction algorithm is formulated that produces high-quality sharp absolute EIT images by combining the D-bar algorithm with subsequent processing by CNN [13]. As a classical deep learning structure, conditional generative adversarial network (CGAN) can improve the detailed characteristics of generated images effectively through adversarial learning between the generator and discriminator [14]. The generator and discriminator are important components in a CGAN, and the difficulty of discrimination increases with improvements in the generation ability. Since the conductivity of the lung is unevenly distributed, the reconstructed images by conjugate gradient algorithm are distorted severely, CGAN is employed to estimate high-quality reconstructed ERT images from severely distorted images.

Inspired by CGAN, a deep learning structure, namely, conditional generative adversarial network with attention mechanism and residual connection (CGAN-AMR) is proposed for lung EIT image reconstruction. To accommodate the highly ill-posed characteristics of the lung EIT imaging process, the generator and discriminator structures are redesigned to obtain high-precision lung images. To solve the problem of loss of details in lung EIT imaging, an attention mechanism is employed to the generator to learn the channel dependencies, focusing on the detail recovery of the lesions, resulting in improvement in the reconstruction accuracy and interpretability of the network. Residual connection is introduced in the discriminator to reduce the difficulty of network learning, which increases the speed of network training and improves the generalization ability of the model to adapt to different lung shapes. In addition, the conductivity distribution of several common lung diseases is collected as the dataset, which is also adopted to assess the performance of the proposed method.

2   Related Work

2.1   Basic Principle of EIT

A typical EIT measurement system is shown in Fig. 1 with 16 evenly distributed electrodes. Current is input through adjacent electrodes and voltage measurements are performed at all other adjacent electrodes, resulting in a total of 208 voltage measurements. The measurements on the boundary of the measured area are affected by the electric field distribution in the area, and the change of the internal medium distribution will cause the current density and electric potential distribution. That is, the voltage measurements measured by the boundary electrode array reflect the internal conductivity of the changed measured area [15].

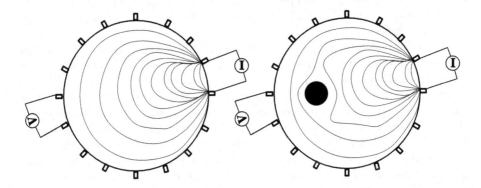

Fig. 1. 16-electrode Lung EIT measurement system.

The EIT problem is a typical nonlinear inverse problem. The goal of the inverse problem in imaging is to estimate the unknown image $\delta\sigma$ from the measured data δu, which is related to through the forward operator S. The forward model of the imaging problem is shown as

$$\delta u = S\delta\sigma + \varepsilon \tag{1}$$

where ε denotes the noise in the measured data. Unlike MRI and CT, EIT imaging has a soft field effects with a nonlinear operator S.

The inverse problem can generally be expressed as a minimum cost function optimization problem as shown

$$\arg\min_{\delta\sigma} F_S(\delta\sigma, \delta u) + \lambda R(\delta\sigma) \tag{2}$$

where F_S denotes the authenticity of the consistency between the reconstructed image $\delta\sigma$ and the measured data δu, and R is a regularization, which improves the image quality by imposing prior knowledge such as smoothness and sparsity. When the disturbance is small, based on the principle of sensitivity coefficient, the principle of sensitivity matrix can be used to reconstruct the conductivity distribution.

The deep learning method is very different from the traditional regularized model fitting method, which can be expressed as

$$R_{learn} = \arg\min \sum_1^N L\{\delta\sigma, R_\theta(\delta u)\} + g(\theta) \tag{3}$$

where R_θ denotes deep learning network with parameter θ; g and L represent the regularization and loss function respectively, and N is the amount of the training samples. In this paper, the conjugate gradient method is used to preprocess the measurement data and obtain the primary mapping image of voltage and conductivity as the input of the CGAN network. The conjugate gradient algorithm is an iterative method for solving symmetric positive definite linear equations. The sensitivity matrix is neither symmetric nor positive definite in the process of EIT image reconstruction. In order to ensure better convergence, it can be expressed as

$$S^T S \delta\sigma = S^T \delta u \tag{4}$$

Assuming that the conductivity is uniformly distributed in the measured area, the initial value of the reconstructed conductivity vector σ_0 is obtained, and calculate in turn.

2.2 Conditional Generative Adversarial Network

Normally generative adversarial network contains a generator and a discriminator, which is described as

$$L_{GAN}(G, D) = E_x[\log D(x)] + E_x[\log(1 - D(G(z)))] \tag{5}$$

where D maximizes $log D(x)$ while G minimizes $log(1 - D(G(z)))$, x is the label image and z is random noise. The output of generator $G(z)$ is a candidate image with the probability distribution of x that maps from the prior distribution of z. The output of discriminator $D(x)$ or $D(G(z))$ is a scalar scoring how close the input is, which is defined as the possibility of input belonging to x.

Conditional GAN is used to reconstruct the field image. It adds extra information y to penalize generator outputs that are beyond the given information. The extra information y can be labels or any other limitations. In practice, y is fed into both discriminator and generator as an additional input layer. The objective function can be rewritten as

$$L_{CGAN}(G, D) = E_{x,y}[\log D(x \mid y)] + E_{x,y}[\log(1 - D(G(z \mid y)))] \tag{6}$$

3   CGAN-AMR: Structure for EIT Image Reconstruction

In the EIT inverse problem, it requires a stable output. Therefore, the input of our generator network is the preprocessed conductivity mapping images obtained by the conjugate gradient algorithm instead of random noise or random dropout. In this paper, the network structure is mainly divided into two modules: data preprocessing module and deep learning module as shown in Fig. 2.

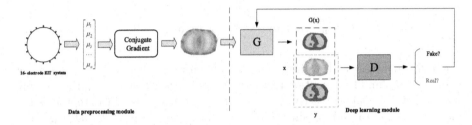

Fig. 2. Schematic of CGAN-AMR.

CGAN-AMR is mainly composed of two modules: the generator G aims to generate the corresponding fake image based on the input image, and the discriminator D is used to distinguish whether the image is the real one or the generated one from G.

Generator: The network structure of the generator is shown in Fig. 3. Since image convolution operations exploit the structural information in medical images well, the U-net is adopted as the generator in this paper. The input and output images of the network are all 3-channel images with a size of 256×256. The generator

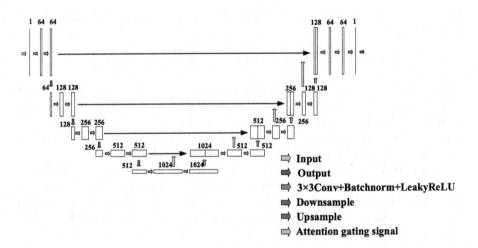

Fig. 3. Structure of generator.

structure consists of down-sampling, up-sampling, and reconstruction processes. In the down-sampling process, four convolution operations are performed and the convolution filter size is 3×3. The image size is halved and the number of channels is doubled. In the up-sampling process, four deconvolution operations are performed and the convolution filter size is 3×3 so that the image size is doubled and the number of channels is halved. In addition, the LeakyReLu activation function and batch normalization are added to sampling operations. As a result, a $1 \times 1 \times 3$ filter is used to reconstruct a clean EIT image.

In order to adapt to the complex feature extraction in the EIT imaging process, based on the structure of U-net, attention mechanism is added to each skip connection to supervise the importance of features to the lesion and obtains the generated image through the global area to obtain more sufficient feature information [16]. The attention module is an effective method for modeling remote dependencies. The attention mechanism is shown as Fig. 4, where X denotes the input image feature, F_{tr} is a convolution kernel operation, the extracted image feature information U is obtained after F_{tr}.

$$z_c = F_{sq}(u_c) = \frac{1}{H \times W} \sum_{i=1}^{H} \sum_{j=1}^{W} u_c(i,j) \tag{7}$$

$$S = F_{ex}(z, W) = \gamma(g(z, W)) = \gamma(W_2 \rho(W_1 z)) \tag{8}$$

$$\hat{X} = F_{scale}(u_c, s_c) = s_c \cdot u_c \tag{9}$$

Firstly, the global averaging process of the feature information of the image is realized by Eq. (7), which shows the numerical distribution of the feature maps of this layer. Secondly, the excitation weight S is obtained to activate each image feature channel according to Eq. (8), in which ρ denotes the ReLU activation function and γ denotes the Sigmoid activation function. Finally, by Eq. (9), the value of each channel is multiplied by the corresponding weight to realize the effective extraction of important channel information.

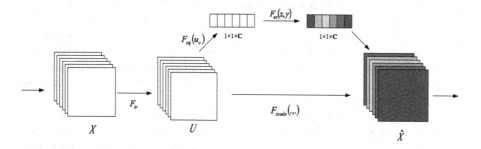

Fig. 4. Schematic of the attention mechanism.

Discriminator: The discriminator network structure designed in this paper consists of nine residual blocks. The nine residual blocks have the same structure. In

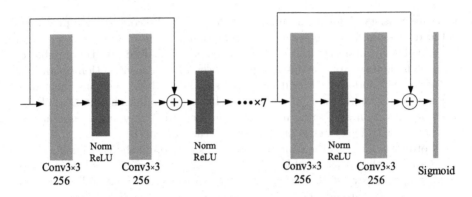

Fig. 5. Structure of discriminator.

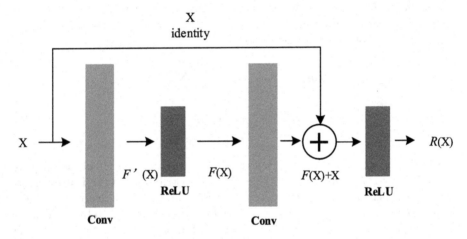

Fig. 6. Structure of residual block.

the last layer, the Sigmoid function is used to output the true probability of each pixel. The structure is as shown in Fig. 5. The input of the discriminator is the ground truth image or the generated image. The output of the discriminator is the probability value that the input image is the ground truth image. When the discriminator determines that the input is ground truth image, the output is 1; on the contrary, when the discriminator determines that the input is generated image, the output is 0.

To make the network training effect more stable, the convolution module is replaced by the residual block, which deepens the numbers of discriminator network layers. A skip connection is established between the various residual blocks, which guarantees the transmission of information to the greatest extent [17].

The basic residual block structure is shown in Fig. 6. X is the input of the residual block and $R(X)$ is the expected output. $F'(X)$ denotes the output of the first layer of the residual block. ω_1, ω_2 denote the weight of the first layer and second layer, respectively. The output of the second-level residual block and the output of the final residual block is expressed as Equation (10) and (11), where ρ denotes the ReLU activation function. Since the "shortcut connection" structure in the residual block is added, the learning goal of the network is the difference between the nonlinear mapping and the congruent mapping, which makes the training of the network more stable and optimized much easier.

$$F(x) = \omega_2 \rho \left(\omega_1 x \right) \tag{10}$$

$$R(x) = \rho(F(x) + x) \tag{11}$$

4   Loss Function

The new loss function is composed of two parts: the confrontation loss and the regression loss, which can be express as

$$L = \lambda_1 \arg \min_G \max_D L_{CGAN}(D, G) + \lambda_2 L_{L1} \tag{12}$$

For lung EIT imaging, the trustworthiness of the reconstructed image is very important, a robust loss L_1 is used to evaluate the difference between ground truth image and the reconstructed image, which can be expressed as

$$L_{L_1} = E_{\delta\hat{\sigma}, y \sim P_r(\delta\hat{\sigma}, y)} \lfloor \|y - \delta\sigma^*\|_1 \rfloor \tag{13}$$

The optimization objective is the process of continuously reducing the loss value L_{L1}. The objective function of the generator and the discriminator with the loss L_{L1} can be expressed as respectively,

$$L_G = \lambda_1 L_{CGAN}(G) + \lambda_2 L_{L_1} \tag{14}$$

and

$$L_D = \lambda_1 L_{CGAN}(D) + \lambda_2 L_{L_1} \tag{15}$$

where λ_1, λ_2 are the weighting terms to balance the adversarial loss and regression loss. The most reasonable λ_1 and λ_2 are selected as $\lambda_1 = 100$ and $\lambda_2 = 10$ through experiment.

5   Experiment

5.1   Dataset

In this paper, three types of lung diseases, i.e., lung tumor, pleural effusion and pulmonary edema, are selected as the reconstructed objects. Based on 2000 CT images of healthy lungs, simulation samples are constructed by COMSOL Multiphysics to manually add lesions of different positions and shapes for network training. According to conductivity characteristics of body tissues, the conductivity of the lung tissue and the lesion tissue are set to 0.0251 S/m and

0.3 S/m respectively. Three types of lung disease models are shown in Table 1. The dataset in CGAN-AMR respectively contained 9000 pairs of samples, and the ratio of the number of lung tumor samples to the number of pleural effusion samples to the number of pulmonary edema samples is 1:1:1. For each disease type, it is a good practice to assign 70% of samples to the training set and 30% to the test set.

Table 1. Details about the dataset.

Type	Model	Quantity
Lung tumor		3000
Pleural effusion		3000
Pulmonary edema		3000

5.2   Evaluation Index

In order to quantitatively analyze the quality of EIT images reconstructed by different networks, PSNR (peak signal to noise ratio) and SSIM (structural similarity) are used to evaluate the effectiveness of the proposed method, which can be expressed as

$$PSNR(i,j) = 10\lg(\frac{255^2 H \cdot W}{\|i-j\|^2}) \tag{16}$$

$$SSIM(i,j) = \frac{(2\mu_i\mu_j + C_1)(2\sigma_{ij} + C_2)}{(\mu_i^2 + \mu_j^2 + C_1)(\sigma_i^2 + \sigma_j^2 + C_2)} \tag{17}$$

where H and W respectively denote the length and width of the images; μ, σ and σ_{ij} are the mean value, standard deviation and covariance of reconstructed image, respectively; i and j denote the pixel indexes of the ground truth images and the reconstructed images; C_i and C_j are constants.

5.3   Results and Analysis

In order to test the performance of the CGAN-AMR, simulation experiments are conducted. The test set contains all the three types of lung lesions in Table 1. Seven specific samples are selected from the test set to evaluate the proposed method (see Fig. 7).

It can be seen from the reconstructed images by three methods in Fig. 7 that the proposed CGAN-AMR has better imaging effects than the CNN method and the traditional CGAN method. The reconstructed images by CNN method has

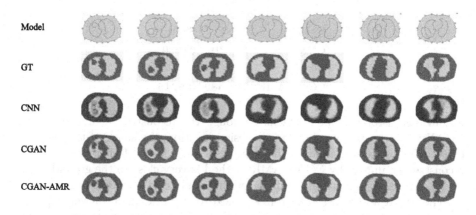

Fig. 7. Reconstruction images.

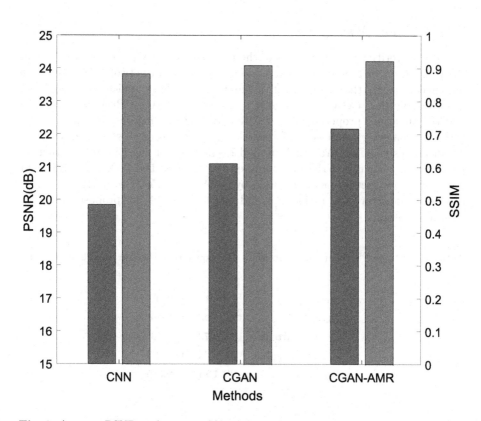

Fig. 8. Average PSNR and average SSIM of reconstructed images by three methods.

Table 2. Average PSNR and average SSIM of reconstructed images by three methods.

Methods	PSNR	SSIM
CNN	19.8391	0.8820
CGAN	21.0940	0.9079
CGAN-AMR	**22.1589**	**0.9215**

no obvious distortion but artifacts; Although the reconstructed images of the CGAN network can contribute to the lesion location, the shape of the lesion is quite different from ground truth image, especially in the last two cases. The reconstructed images by CGAN-AMR method is very close to the ground truth image, and the image quality is higher with complete margin retention.

The PSNR and SSIM indicators are used to evaluate quantitatively the difference between the images generated and the groud truth images in the EIT reconstruction task, respectively. The reconstructed images by the three methods in the EIT reconstruction task are quantitatively evaluated using PSNR and SSIM, respectively. The results are shown in Table 2 and Fig. 8.

5.4   Ablation Experiment

In order to verify the effectiveness of the improved CGAN network structure on the problem of EIT image reconstruction. Four types of network structures are designed to verify the lifting effect of the added modules, which are traditional CGAN structure, CGAN + Attention structure, CGAN+ Residual structure and the structure proposed. The PSNR and SSIM indicators are employed to verify the imaging results of different network structures as shown in Table 3, which proves that the proposed method is effective for the EIT reconstruction problem. As shown in Table 3, compared the CGAN, the PSNR and SSIM of the proposed structure have increased by 5.05% and 1.50%, respectively. These results demonstrate the rationality and effectiveness of the proposed network structure design.

Table 3. Average PSNR and average SSIM of reconstructed images by different structures

Methods	PSNR	SSIM
CGAN	21.0940	0.9079
CGAN+Attention	21.2673	0.9189
CGAN+Residual	21.1724	0.9105
Proposed	**22.1589**	**0.9215**

6   Conclusion

In this paper, a novel conditional generative adversarial network with attention mechanism and residual connection (CGAN-AMR) is proposed for lung EIT imaging. In addition, a lung diseases simulation dataset is produced to evaluate the performance of the proposed method. Comprehensive experiments are conducted to evaluate the effectiveness and generality of the proposed method, which show that the proposed method obtains the best visual quality and higher objective indexes compared with CNN and CGAN networks. The lesions can be identified with complete boundaries and exact location. That means the proposed method is suitable for lung EIT imaging with the potential for lung disease detection. Besides, how reconstructing multi-conductivity distribution samples based on the CGAN-AMR is another focus that will be continuously explored in future work.

References

1. Fan, Y., Ying, L.: Solving electrical impedance tomography with deep learning. J. Comput. Phys. **404**, 109–119 (2020)
2. Wang, Z., Yue, S., Liu, X.: An iterative linear back-projection algorithm for electrical impedance tomography. In: 2018 13th World Congress on Intelligent Control and Automation (WCICA), pp. 484–489. IEEE, Changsha (2018)
3. Li, F., Dong, F., Tan, C.: Landweber iterative image reconstruction method incorporated deep learning for electrical resistance tomography. IEEE Trans. Instrum. Meas. **70**, 1–11 (2020)
4. Grootveld, C.J., Segal, A., Scarlett, B.: Regularized modified Newton-Raphson technique applied to electrical impedance tomography. Int. J. Imaging Syst. Technol. **9**(1), 60–65 (1998)
5. Vauhkonen, M., Vadász, D., Karjalainen, P.A., et al.: Tikhonov regularization and prior information in electrical impedance tomography. IEEE Trans. Med. Imaging **17**(2), 285–293 (1998)
6. Zemouri, R., Zerhouni, N., Racoceanu, D.: Deep learning in the biomedical applications: recent and future status. Appl. Sci. **9**(8), 15–26 (2019)
7. Yamashita, R., Nishio, M., Do, R.K.G., Togashi, K.: Convolutional neural networks: an overview and application in radiology. Insights Imaging **9**(4), 611–629 (2018). https://doi.org/10.1007/s13244-018-0639-9
8. Lee, D., Yoo, J., Tak, S., et al.: Deep residual learning for accelerated MRI using magnitude and phase networks. IEEE Trans. Biomed. Eng. **65**(9), 1985–1995 (2018)
9. Hamghalam, M., Lei, B., Wang, T.: High tissue contrast MRI synthesis using multi-stage attention-GAN for segmentation. In: Proceedings of the AAAI Conference on Artificial Intelligence, pp. 4067–4074. AAAI, New York (2020)
10. Zheng, J., Peng, L.: An autoencoder-based image reconstruction for electrical capacitance tomography. IEEE Sens. J. **18**(13), 5464–5474 (2018)
11. Hu, D., Lu, K., Yang, Y.: Image reconstruction for electrical impedance tomography based on spatial invariant feature maps and convolutional neural network. In: 2019 IEEE International Conference on Imaging Systems and Techniques (IST), pp. 1–6. IEEE, Abu Dhabi (2019)

12. Tan, C., Lv, S., Dong, F., et al.: Image reconstruction based on convolutional neural network for electrical resistance tomography. IEEE Sens. J. **19**(1), 196–204 (2018)
13. Hamilton, S.J., Hauptmann, A.: Deep D-bar: Real-time electrical impedance tomography imaging with deep neural networks. IEEE Trans. Med. Imaging **37**(10), 2367–2377 (2018)
14. Creswell, A., White, T., Dumoulin, V., et al.: Generative adversarial networks: an overview. IEEE Signal Process. Mag. **35**(1), 53–65 (2018)
15. Cheney, M., Isaacson, D., Newell, J.C.: Electrical impedance tomography. SIAM Rev. **41**(1), 85–101 (1999)
16. Hu, J., Shen, L., Sun, G.: Squeeze-and-excitation networks. In: Proceedings of the IEEE Conference on Computer Vision and Pattern Recognition, pp. 7132–7141. IEEE, Salt Lake City (2018)
17. He, K, Zhang, X, Ren, S, et al.: Deep residual learning for image recognition. In: Proceedings of the IEEE Conference on Computer Vision and Pattern Rcognition, pp. 770–778. IEEE, Las Vegas (2016)

Snake Robot Motion Planning Based on Improved Depth Deterministic Policy Gradient

Xianlin Liu[1], Jianming Wang[2], and Yukuan Sun[3(✉)]

[1] Software College, Tiangong University, Tianjin 300387, China
[2] School of Computer Science and Technology, Tiangong University, Tianjin 300387, China
wangjianming@tiangong.edu.cn
[3] Center for Engineering Intership and Training, Tiangong University, Tianjin 300387, China
sunyukuan@tiangong.edu.cn

Abstract. Due to the complexity of modular mechanism control, the motion planning of snake robots in an unfamiliar complex environment is a long-standing issue. We propose an improved deep deterministic policy gradient (DDPG) algorithm to plan the path with the shortest time and the least collisions. Traditional robot DDPG can not make full use of previous states to make decisions. This paper uses LSTM, learns all previous hidden states through memory and reasoning, and distinguishes the importance of features through the Self-Attention mechanism, which reduces the impact of useless features on decision-making and improves decision-making accuracy. In addition, the reward function is optimized to make the snake robot reach the target point faster. Finally, we do experiments in the simulation environment. The results show that the algorithm can speed up the convergence speed of the DDPG and reduce the path planning time and collision times of the snake robot.

Keywords: Snake Robot · DDPG · LSTM · Self-Attention

1 Introduction

Robot technology has been applied more and more successfully in industry, national defence, civil and agricultural fields, and people's work efficiency and quality have been further improved. In the past decades, a variety of robots have been developed. Among them, the snake robot has attracted much attention because of its wide application in rescue, medical and other fields. However, its high degree of freedom (DOF) modular joints help them change shape and navigate motion in a highly chaotic environment. However, this modularization makes the navigation and obstacle avoidance of snake robots in the environment a very challenging task.

The early research of snake robots mainly used the biological knowledge of the true serpents in nature, and the concept of snake robot was first proposed by Hirose [1] in

This work was supported by The National Natural Science Foundation of China (62072335) and The Tianjin Science and Technology Program (19PTZWHZ00020).

S. Yang and S. Islam (Eds.): APWeb-WAIM 2022 Workshops, CCIS 1784, pp. 151–162, 2023.
https://doi.org/10.1007/978-981-99-1354-1_14

1993. A lot of research has been carried out on snake motion (gait design), mainly using the method based on sine curve [2], dynamics [3], and central pattern generator [4]. These algorithms lack interaction with the environment to adapt to different terrain and climate. Therefore, it is pretty difficult for the model-based method to adaptive control the robot in complex and demanding environments.

In addition, in the aspect of obstacle-assisted snake robots [7], Sartoretti et al. [8] studied the A3C algorithm for snake robot control. Jaryani et al. [9] proposed autonomous navigation and obstacle avoidance model of a planar snake robot based on speed heading joint custody, which uses a virtual target to generate control commands so that the snake robot can track the moving virtual target while avoiding the obstacle of the virtual target. On the other hand, Yang et al. used the improved rapid exploration random tree method to generate feasible paths in [10] and proposed a time-varying line-of-sight (LOS) guidance law and a cubic spline interpolation (CSI) path planning method in [11]. However, these methods are not suitable for crowded scenes with obstacles and are easy to fall into deadlock.

Recent work on path planning using reinforcement learning (RL) [5] & [6] shows good results of robot adaptive motion planning. In reference [7], a DRL-based method is proposed, which relies on a monocular camera for perception and vision-based obstacle avoidance from the original sensory data. We propose a new snake robot motion planning algorithm based on depth RL. In this technology, we use the improved deep DPG network (DDPG) for motion planning. DDPG is a model-free RL algorithm that allows neural networks to operate in high-dimensional visual state space. A method of calculating gradient is designed to solve the problem of using a standard gradient, and satisfactory results are achieved in various tasks [14].

The main goal of our method is to reach the target point in the shortest time without colliding with obstacles. We regard the angle and distance between the snake's forward direction and the target point as the state space of the snake robot itself. In order to make full use of the features in the environment, explore and interact with the location environment by using the data-driven method, and improve the efficiency of the RL framework, we use image input as the environmental state of the snake robot interacting with the environment, which is helpful to learn complex and hierarchical abstract feature representation. The main contributions are as follows:

- A new network structure based on DDPG is proposed. The DDPG network structure is optimized by using the "memory" ability of LSTM and the "attention" ability of Self-Attention, and the motion planning of the snake robot is realized based on this method.
- A reward function suitable for obstacle avoidance and path planning of the snake robot is designed. The goal is to reach the target point in the shortest time without collision with obstacles.
- In order to make full use of the information in the environment for path planning, the snake robot is designed to integrate its own state with the environmental state as the current state of deep reinforcement learning.

2 Related Works

2.1 Deep Deterministic Policy Gradient (DDPG)

The DDPG [13] is on account of the Actor-Critic architecture and combines the advantages of DQN [14] and DPG [15], which not only solves the problem of low efficiency when DQN processes continuous action space but also introduces a deep neural network for end-to-end operation. DDPG adopts another method to update the Q function parameters as the target:

$$\theta^{'} = \tau\theta + (1 - \tau)\theta^{'} \tag{1}$$

where $\theta^{'}$ and θ are the parameters of Q function as target and Q process to be optimized respectively, τ is a constant and $\tau \ll 1$. That is, $\theta^{'}$ and θ are updated at each iteration. The randomization policy for sampling in DPG is materialized. The specific formula is as follows:

$$\mu^{'}(s_t) = \mu\big(s_t|\theta_t^{\mu}\big) + N \tag{2}$$

where $\mu^{'}$ is the randomization policy for sampling, μ is the deterministic policy to be optimized, and N is the random noise. This optimizes μ indirectly while optimizing $\mu^{'}$.

In the process of path planning, the algorithm uses a random process of μ and random OU(Ornstein Uhlenbeck) noise generation according to the current online strategy. The robot takes its own state data and the observed nearby environment information as state s. The actor-network outputs the action a that the robot needs to perform according to s and random process. The robot then obtains the corresponding reward r from the environment by performing the action. Depending on the status and action a, the critical network produces the Q value as a valuation of the action and continuously adjusts its value function. The network of actors will continually improve the action strategy by value Q. The target network among actors and critics is mainly used to update the subsequent process. Figure 1 shows the structure of the DDPG algorithm.

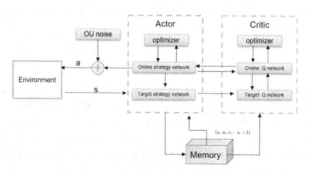

Fig. 1. DDPG network structure

As a mainstream deep reinforcement learning algorithm, DDPG has been widely used in the field of robot motion planning. Because the algorithm adopts the continuous state space and the space of action, it fits the planning of the snake robot's motion and shows great potential in a complex environment.

2.2 Long Short-Term Memory (LSTM)

A recurrent neural network (RNN) is a neural network that processes time series data. RNN takes time series data as input and gives the network the ability to remember previous states. However, due to the disappearance of the gradient, there can only be short-term memory. LSTM [12] is an improvement on the basis of RNN. It solves the problem of combining short-term memory with long-term memory by using a gating mechanism so that it can learn long-term dependent information. Compared with RNN, LSTM introduces three gating mechanisms, namely forget gate, input gate, and output gate. Through the gating mechanism, timely judge and adjust the information at every moment and update it.

LSTM can fully use the previous information and is fit for the processing and prediction of applications with long time series. Planning the trajectory of the Snake robot is a typical long-sequence decision problem. When the DDPG algorithm is used to plan the path of the snake robot, It can only rely on the state information of the snake robot at the current moment, resulting in a too tortuous and inefficient path. In this paper, the LSTM is introduced in the proposed DDPG network structure with reference to the overall use of the past and present state of snake robots for trajectory planning.

2.3 Self-attention

Self-Attention [17] evolved from attention. The basic idea is to calculate the weights of different parts and then calculate the weighted sum. The final effect is to pay varying degrees of attention to different parts of the input feature. The difference is that the attention mechanism occurs between the target element query and all elements in the source. Self-Attention refers to the attention mechanism between elements inside the source or between elements inside the target.

Self-Attention does not contain any RNN and CNN structures, which can solve the problem of long-range dependence of sequences. In addition, matrix calculation can be very fast because of parallelization. The attention feature vector h is calculated by Eq. 3, where Q, K and V are three vectors respectively obtained from the X-linear mapping of input features, and d_k is the row dimension of K.

$$Attention(Q, K, V) = softmax\left(\frac{K^T Q}{\sqrt{d_k}}\right) V \qquad (3)$$

Because our snake robot uses images as environmental information, the amount of information is huge for long-sequence path planning. Although LSTM alleviates the computational burden of the network, it does not distinguish the importance of features, so some features with low importance may also have a great impact on decision-making; too many useless features will reduce the efficiency of the snake robot path planning. In this paper, Self-Attention is introduced into the DDPG network structure to selectively use environmental information to plan the path of the snake robot.

3 Methodology

We propose an improved motion planning method for the DDPG snake robot based on RL. The main goal of the agent is to use the snake robot's serpentine sinuous gait to reach the target point from the current position in the shortest time step without collision. Because the traditional DDPG algorithm may lead to the too tortuous path when applied to path planning, LSTM and Self-Attention are introduced to improve and optimize the DDPG network structure, and the reward function is improved on the basis of [20] to speed up the network training.

3.1 Network Structure

Each layer in the network structure of the original DDPG algorithm is a fully connected layer. When using this algorithm to plan the path of the robot, due to the limited environment that the snake robot can observe and the lack of previous state information, the path planning can only rely on the limited environment information and current state information obtained by the snake robot, and the path planning is very chaotic. Using the memory ability of LSTM, we should introduce it into the learning process of the snake robot. The hidden vector obtained by LSTM already contains the rich feature information of the previous frame of video, but LSTM can only generate a fixed length vector for input and will not distinguish the importance of information. All features have the same weight, which may cause large decision errors. Therefore, we introduce Self-Attention into the learning process of the snake robot. The improved network structure is shown in Fig. 2.

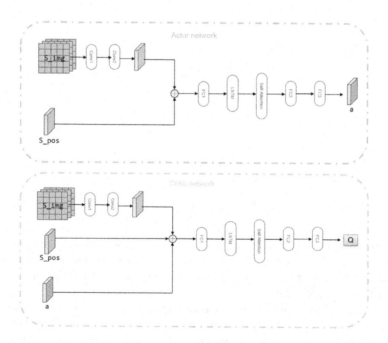

Fig. 2. Improved network structure

DDPG uses an Actor-Critic frame structure, and the structure of its target network is the same as that of online network, both of which adopt the above LSTM structure and

Self-Attention structure. Among them, the combination of image state and position state is used as the input of the model; that is, the image information collected by the snake head camera is feature extracted through two convolution layers (conv1 and conv2); the extracted features are feature stitched with the position state introduced in 3.2, the Critic network needs to splice the actions output by the actor-network. And then input to LSTM through a full connection layer FC1 for memory reasoning and learning. The feature vector obtained is weighted by the Self-Attention module. Finally, the action is predicted through the last two full connection layers. See the experimental module in Sect. 4 for the specific network parameter settings.

3.2 State and Action Space

State space is the feedback of the snake robot to the environment and is the premise of the snake robot to make decisions and select actions. In order to make full use of environmental information for learning, we propose a method combining image information (S_img) with its own state (S_pos) on the basis of [20], where S_pos adopts the definition of 4-tuple proposed, as shown in Fig. 3 and Eq. 4, to describe the relative state of the robot and the environment.

$$s\_pos =< \varphi_1, d_1, \varphi_2, d_2 > \tag{4}$$

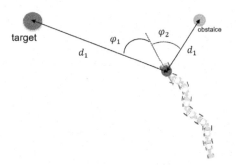

Fig. 3. Status of the robot

Where φ_1 is the angle between the snake robot and the target; d_1 is the distance between the robot and the target, which is used to evaluate whether the snake robot is approaching the target position; φ_2 and d_2 are the angle and distance between the snake robot and the obstacle (nearest).

The image is easily affected by the environment in the process of acquisition, transmission and conversion, which is represented as noise in the image. These noises will reduce the image quality or interfere with our extraction of the originally desired image information. Therefore, on the basis of obtaining the image collected by the snake head camera, we perform Gaussian filtering on the image to remove the noise interference in these images. In order to speed up the network training, we use $32 \times 32 \times 3$ image information as our image state S_img.

We choose to adopt the serpenoid model proposed by Professor Hirose,

$$\theta = \alpha_t \sin(\omega t + i\beta) + \gamma_t \tag{5}$$

where θ is the joint angle; ω is the time-frequency that determines the speed of the wave; β is the module phase; i is the robot joint index.

In order to effectively reduce the dimension of action space and speed up the convergence of the network, we fix ω and β constant and choose a $= \Delta\alpha$, Where $\Delta\alpha$ is the amplitude transformation of serpenoid curve. Specifically, it is a constant amplitude A under the premise of increment $\Delta\alpha$ defined as amplitude $-\xi < \Delta\alpha < \xi$, where ξ is a constant. The joint angles θ are calculated by Eq. 6.

$$\theta = (A + \Delta\alpha)\sin(\omega t + i\beta) + \gamma_t \tag{6}$$

3.3 Reward

We hope that the snake robot to be able to reach the target position in the shortest time and avoid obstacles. However, in real life, the snake robot may slide when winding, and this uncontrollable sliding will cause collisions when the snake robot moves. In order to make the snake robot reach the target point more quickly and avoid the influence caused by sliding, we added a reward limiting sliding on the basis of the reward proposed in [20] and made a new design for the reward limiting the redundant movement of the robot. Rewards for this paper are shown below:

$$r_t = r_1 + r_2 + r_3 + r_4 \tag{7}$$

where:

$$r_1 = -k_1 d_1, \tag{8}$$

$$r_2 = \begin{cases} 0 \\ \frac{-k_2}{d_2+1}, d_2 < th_2 \end{cases} \tag{9}$$

$$r_3 = \begin{cases} 0 \\ k_3, d_1 < th_1 \end{cases} \tag{10}$$

$$r_4 = \begin{cases} 0 \\ -k_4, \varepsilon > th_4 \end{cases} \tag{11}$$

Among them, reward r_1 is used to punish any redundant action that does not advance toward the goal so as to encourage the snake robot to reach the specified target position in the shortest time; Reward r_2 is used to avoid collision with obstacles in the process of movement; Reward r_3 is the final reward when the goal is achieved; Reward r_4 is used for the punishment of sliding. In the reward function reward feedback, we use the relationship between the joint position of the snake robot and the distance between the obstacle and the target in the environment to determine whether to touch the obstacle or reach the target point. When the robot moves in the actual environment, it will inevitably slide; that is, the joint position of the snake robot can not reach the ideal state, which will affect the judgment of its own state such as distance, included angle, and so on. In order to solve the above problems, we add a laser radar to the snake head position and compare the joint position detected by the radar at each step with the joint position calculated by the forward kinematics. If the error $\varepsilon_i > th_4$, the joint position of the snake robot is corrected, and $\gamma_t = \frac{\gamma_k}{d_2+1}$ in Eq. 6, where th_4 is the deviation threshold we assume. In our experiment, we set it as $th_4 = 0.05$, and γ_k is a constant, $k_1 = 0.01, k_2 = 0.1, k_3 = 200$, $k_4 = 0.02$, th_1 and th_2 have different values in different scenes. See the experiment in Sect. 4 for specific settings.

4   Experimental Results and Analysis

We use the simulation platform "Webots" for experimental evaluation. Webots is an excellent open source multiplatform robot simulation software which provides a complete development environment for robot modelling, programming, and simulation. By importing predesigned 3D-CAD models, various robots can be created flexibly.

Fig. 4. Snake robot simulation model

In our experimental evaluation, the snake-shaped robot used consists of eight joints (as shown in Fig. 4). Use joint angle θ to control each actuator because robots interact with targets and obstacles and θ changes with the potential system uncertainty. In particular, as shown in Fig. 3, Firstly, the image information of the current moment is collected, and the relative state information between the snake robot and the environment is dynamically estimated. Then, the learned RL strategy is calculated to obtain an action. Finally, the Angle change of the ideal actuator can be calculated by Eq. 6 and transmitted to the snake robot controller.

For the selection of experimental scenarios, we use two environments; Scene 1 is a 4 m × 6 m small scene intensive obstacle environment; Scene 2 is a 20 m × 12 m large scene sparse obstacle environment. We use the above-modified network as the basic network of online network and target network for DDPG. Conv1 in the neural network uses 16 kernels of size 3 × 3, and the step size is 2; Conv2 uses 32 kernels of size 3 × 3, and the step size is 2; The output of both uses the ReLU activation function, in which the full connection layers FC1 and FC2 use 150 hidden units. Finally, an output layer is added, and the size of the output layer corresponds to the number of operations available (in this example, the FC3 layer of the actor network and the critic network output 3 and 1 values). In addition, our optimizer uses "Adam" and the activation function in the output layer makes use of a linear activation function. Table 1 shows all the super parameters of our network.

Table 1. Network parameters

Parameter	Value
Activation function for hidden layers	relu
Activation function for output layer	linear
Optimizer	Adam
Learning rate for actor network	1e-3
Learning rate for criter network	2e-3
Initial epsilon	1
Final epsilon	0.01
Update epsilon (in steps)	4
Target network update frequency	16
Memory size	1000
Batch size	64

4.1 Scene 1: Dense Scene

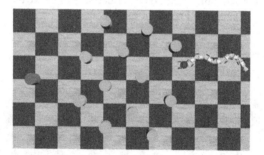

Fig. 5. Simulation diagram of the dense scene

Our dense scene is shown in Fig. 5. We set th_1 and th_2 to 15 and 12. Train the model and compare it with DDPG_LSTM designed in [18] and TD3 (The improvement of DDPG) [19]. Figure 6 shows the changes in data collected in the training process after model convergence. Figure 6(a) shows the comparison of loss change curves of the actor part, and Fig. 6(b) shows the comparison of loss change curves of the critic part. Observing the two figures, we can find that both parts of our model can converge and have advantages compared with the other two algorithms, which shows that the model designed in this paper is reasonable and feasible.

Table 2 shows the average quantitative analysis results of five repeated simulations by different methods. We can draw the following conclusions: 1) the average routing

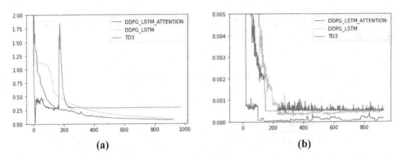

(a)   (b)

Fig. 6. Loss: (a) is the loss comparison of actor, (b) is the loss comparison of critic

Table 2. Number and time of collisions in dense scenes

Average Item	Our	DDPG_LSTM	TD3	Improvement
Routing Time	3.52	5.04	5.28	30% ~ 33%
Collisions	2	5	6	60% ~ 67%

time of the snake robot using our method is much less than DDPG_LSTM [18] and TD3 [19]. Under the same target, its navigation efficiency is improved by more than 30%. 2) our method only has two collisions, while DDPG_LSTM and TD3 have more collisions, which shows that this method can effectively improve the obstacle avoidance performance of the snake-shaped robot in such scenes.

4.2   Scene 2: Sparse Scene

Fig. 7. Sparse large scene simulation

Figure 7 shows some results of the snake robot moving from start to target. The pink track shows the overall motion of the snake-shaped robot, in which the robot can successfully avoid obstacles and reach the target point. th_1 and th_2 are set to 18 and 20.

Table 3. Comparison of collision times and time in sparse scenes

Average Item	Our	CBRL	MFRL	DLDC
Routing Time	22.95	23.10	52.41	52.95
Collisions	0	2	10	8

Table 3 shows the average quantitative analysis results of five repeated simulations by different methods. We can draw the following conclusions, 1) compared with MFRL [13] and DLDC [8], the average routing time of the snake robot using our method is effectively reduced, and it is also slightly improved compared with CBRL [20] method. Under the same target, its navigation efficiency is more than 50% higher than MFRL and DLDC. 2) compared with the two collisions of CBRL, our method can almost achieve no collision in this scene where the distance between obstacles is large. There are more collisions between MFRL and DLDC, which shows that this method can effectively improve the obstacle avoidance performance of the snake robot in such scenes.

5   Conclusion

Due to the complexity of the snake robot control task, the model-based method has poor robustness in the adaptive control of the robot in an extremely complex environment. We propose a scheme based on improved DDPG to optimize the control of the snake robot; Due to the "memory" ability of LSTM and the attention mechanism's ability to pay attention to important features, we optimize the network structure of DDPG by introducing LSTM and Self-Attention. By designing and improving the reasonable reward function, the path planning model of the snake robot can be quickly trained. This can effectively improve the detection efficiency of the snake robot in complex and unknown environments, so that the snake robot can avoid obstacles and reach the target position in a short time. For the problem of snake robot sliding during motion, we use radar point cloud information to correct the snake robot's own state and enhance the robustness of robot control. However, the algorithm in this paper can effectively avoid obstacles and reach the target position in an environment containing only static obstacles. Dynamic obstacles are also an important factor to be considered in some real scenes. How to effectively avoid dynamic obstacles in snake robot path planning is a future research direction.

Acknowledgement. This work was supported by The National Natural Science Foundation of China (62072335) and The Tianjin Science and Technology Program (19PTZWHZ00020).

References

1. Hirose, S., Cave, P., Goulden, C.: Biologically Inspired Robots: Snake-Like Locomotors and Manipulators. Oxford University Press, Oxford (1993)

2. Tony, O.: Biologically Inspired Robots: Snake-Like Locomotors and Manipulators by Shigeo Hirose Oxford University Press, Oxford (1993). 220 pages, incl. index (£40)." Robotica **12**, 282 (1994)

3. Miller, G.S.P.: The motion dynamics of snakes and worms. In: Proceedings of the 15th Annual Conference on Computer Graphics and Interactive Techniques (1988)

4. Bing, Z., et al.: Towards autonomous locomotion: CPG-based control of smooth 3D slithering gait transition of a snake-like robot. Bioinspiration & Biomimetics **12**(3), 035001 (2017)

5. Li, D., et al.: 2D obstacle avoidance method for snake robot based on modified artificial potential field. In: 2017 IEEE International Conference on Unmanned Systems (ICUS), pp. 358–363 (2017)

6. Hong, J., et al.: Semantically-aware strategies for stereo-visual robotic obstacle avoidance. In: 2021 IEEE International Conference on Robotics and Automation (ICRA), pp. 2450–2456 (2021)

7. Kano, T., et al.: Local reflexive mechanisms essential for snakes' scaffold-based locomotion. Bioinspiration Biomimetics **7**(4), 046008 (2012)

8. Sartoretti, G., et al.: Distributed learning of decentralized control policies for articulated mobile robots. IEEE Trans. Robot. **35**, 1109–1122 (2019)

9. Haghshenas-Jaryani, M., Sevil. H.E.: Autonomous navigation and obstacle avoidance of a snake robot with combined velocity-heading control. In: 2020 IEEE/RSJ International Conference on Intelligent Robots and Systems (IROS), pp. 7507–7512 (2020)

10. Yang, W., et al.: Perception-aware path finding and following of snake robot in unknown environment. In: 2020 IEEE/RSJ International Conference on Intelligent Robots and Systems (IROS), pp. 5925–5930 (2020)

11. Yang, W., et al.: Spline based curve path following of underactuated snake robots. In: 2019 International Conference on Robotics and Automation (ICRA), pp. 5352–5358 (2019)

12. Greff, K., et al.: LSTM: a search space odyssey. IEEE Trans. Neural Netw. Learn. Syst. **28**, 2222–2232 (2017)

13. Arulkumaran, K., et al.: Deep reinforcement learning: a brief survey. IEEE Signal Process. Mag. **34**, 26–38 (2017)

14. Mnih, V., et al.: Human-level control through deep reinforcement learning. Nature **518**, 529–533 (2015)

15. Silver, D., et al.: Deterministic policy gradient algorithms. In: ICML (2014)

16. Liu, X., et al:. Learning to locomote with artificial neural-network and CPG-based control in a soft snake robot. In: 2020 IEEE/RSJ International Conference on Intelligent Robots and Systems (IROS), pp. 7758–7765 (2020)

17. Vaswani, A., et al.: Attention is all you need. ArXiv abs/1706.03762 (2017)

18. Gong, H., et al.: Efficient path planning for mobile robot based on deep deterministic policy gradient. Sensors (Basel, Switzerland) **22**, 3579 (2022)

19. Fujimoto, S., et al.: Addressing function approximation error in actor-critic methods. ArXiv abs/1802.09477 (2018)

20. Jia, Y., Ma, S.: A coach-based bayesian reinforcement learning method for snake robot control. IEEE Robot. Autom. Lett. **6**, 2319–2326 (2021)

Global Path Planning for Multi-objective UAV-Assisted Sensor Data Collection: A DRL Approach

Rongtao Zhang[1,2]($\boxtimes$), Jie Hao[1,2], Ran Wang[1], Hui Wang[1], Hai Deng[1], and Shifang Lu[3]

[1] College of Computer Science and Technology, Nanjing University of Aeronautics and Astronautics, Nanjing, China
{zrt1997,haojie,wangran,wanghui9727,denghai}@nuaa.edu.cn
[2] Collaborative Innovation Center of Novel Software Technology and Industrialization, Nanjing, China
[3] Cetc Information Technology System Co., Ltd., Shanghai, China

Abstract. With the rapid development of the Internet of Things (IOT), the amount of business data carried by wireless sensor networks (WSNs) has exploded. Unmanned aerial vehicle (UAV)-assisted sensor data collection is effective and environmentally friendly. A UAV should determine which sensor nodes to visit, and the order in which to visit them according to the target requirements. Due to UAV energy and data-collection demands, a mission can have more than one goal. This study investigates the multi-objective global path planning of UAV-assisted sensor data collection (SDC). It is modeled as a multi-objective optimization problem (MOP) to maximize the amount of data collected and minimize UAV flight time. Based on a deep reinforcement learning (DRL) framework, we decompose the MOP into a series of subproblems, model them as neural networks and use an actor-critic algorithm and modified pointer network to solve each subproblem. Then the Pareto front of the path planning solution under the above constraints can be obtained through the forward propagation of the neural network. Experimental results show that the proposed method is superior to a traditional multi-objective evolutionary algorithms in terms of convergence, diversity of solutions, and time complexity. In addtion, the proposed method can be directly applied to the situation in which the number of sensors changes without retraining, and it has better robustness.

Keywords: Pointer network · Multi objective optimization · Deep reinforcement learning

---

This work is supported in part by the National Key R&D Program of China under Grant 2019YFB2102000, and in part by the Collaborative Innovation Center of Novel Software Technology and Industrialization.

S. Yang and S. Islam (Eds.): APWeb-WAIM 2022 Workshops, CCIS 1784, pp. 163–174, 2023.
https://doi.org/10.1007/978-981-99-1354-1_15

1   Introduction

Wireless sensor networks (WSNs) are of great significance in business and life. WSN technology has penetrated fields such as industrial production, environmental protection, resource investigation, medical diagnosis, bioengineering, space development, ocean exploration, and even the protection of cultural relics. There are both static and mobile methods of data collection for WSNs, respectively accomplished by uploading sensor data to a data center by multi-hop routing [1] and through a movable collector.

UAV-assisted sensor data collection (SDC) is an economical and efficient method that will play a pivotal role in next-generation wireless communication, with ubiquitous connections and wide coverage [2,3]. UAV-assisted mobile data collection has great advantages over SDC. It requires no ad hoc networks, whose cost-versus-benefit can be quite unbalanced in extreme natural environments. A UAV directly collects the generated data of ground equipment, which can eliminate issues such as route establishment and maintenance, traffic hotspot issues, radio frequency interference, and transmission conflicts. Furthermore, UAV-assisted SDC can respond quickly if an ad hoc network is damaged.

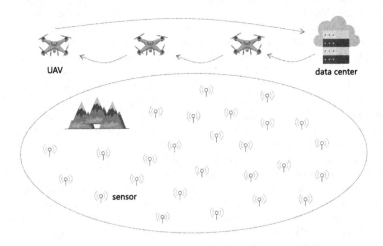

Fig. 1. Problem scenario : the UAV collects the data from the sensors and returns to the data center.

UAV-assisted SDC has some key issues to be addressed. Consider the scenario in Fig. 1. There are a series of sensor nodes in the two-dimensional plane. The UAV should travel to different sensor nodes, hover over them to collect data, and send the data to the data center. However, the sensor data in the region are redundant, and not all sensor nodes must be accessed, so the number of sensors serviced by the UAV may vary between missions. In addition, due to energy constraints and mission requirements, a UAV should collect as much data as possible while minimizing flight time. Multiple targets in a scene may conflict with each other. To plan a feasible multi-objective UAV-assisted SDC global path in various conditions is a complex problem.

More than one objective is considered in the process of UAV-assisted SDC: the amount of data collected and the flight time of the UAV, including the flight distance and the amount of data in the sensor nodes. These two aspects conflict, because more collected data means more flight time. In this case, we expect to get a set of Pareto solutions to complete this task. It is a pity that traditional multi-objective evolutionary algorithms (MOEAs), which can obtain good solutions to multi-objective problems, cannot meet the time and robustness requirements of UAV-assisted SDC. We propose a deep reinforcement learning (DRL) method to solve this problem. The main contributions of this work are as follows:

- In global path planning for the SDC problem, we seek to maximize the amount of data collected and minimize the UAV flight time, including traveling and hovering. We model this as a multi-objective optimization problem (MOP), and propose a framework of global path planning for the multi-objective UAV-assisted SDC problem, which utilizes a neighborhood-based parameter transfer strategy and actor-critic algorithm. We believe this is the first use of DRL to solve global path planning for a multi-objective SDC problem and obtain a Pareto optimal solution.
- Experiments show that the proposed method has superior convergence, diversity, and time complexity compared with the traditional MOP algorithm. When the locations of sensor clusters change, traditional MOEAs must be computed from scratch due to a variable distance matrix, which is unacceptable in practice. Our method is robust to scenario change. Once trained, the model can be applied to new problem situations.

The remainder of this paper is organized as follows. Section 2 introduces some related work. Section 3 describes the system model and problem form. The solution method is shown in Sect. 4. Section 5 shows the solution effect of the model, and Sect. 6 summarizes the study.

2   Related Work

Much of the recent literature on UAV-assisted SDC focuses on a single target. A UAV-assisted Internet of Things (IoT) data acquisition system was proposed to minimize the task completion time [4], using a twin-delayed deep deterministic policy gradient method to address practical urban line-of-sight channel modeling and the availability of imperfect channel state information problems and visit as many ground nodes as possible in a limited time. The UAV sensor collection problem was regarded as a dynamic optimization problem to minimize the data collection time in a linear network [5]. Some work has considered multiple objectives while combining multiple factors as one objective. The authors of [6] presented a framework to cooperatively optimize the communication throughput and propulsion energy consumption of a UAV. An IoT network supporting multiple UAVs was designed to minimize the deployment cost under conditions of limited time and energy [7]. However, we need to trade off between multiple objectives according to specific demands in practical application.

DRL-based multiobjective optimization algorithm (DRL-MOA), an end-to-end framework based on DRL, was proposed to solve the MOP [8] with better generalization ability, faster computing speed, and better convergence and diversity of solutions than the traditional MOEA. A pointer network of deep learning was used to solve the global path planning problem of UAVs collecting sensor data, but the tradeoff between multiple objectives was not considered [9].

3   System Model

We focus on global path planning for data collection using a cluster head node (CHN). Due to the large number of sensors, an algorithm can be used to cluster randomly distributed sensor nodes. The center coordinate of the obtained cluster is used as the CHN. There are many mature clustering algorithms, and they are not the focus of our research. We directly use a random distribution to represent the CHNs after clustering. The task of the UAV is to collect the data of CHNs, one by one, until the number of accessed CHNs meets the requirement. Our goal is to maximize the amount of data collected and minimize the UAV flight time. The UAV starts from one CHN and successively accesses the remaining CHNs until the mission requirements are met. Let $C = \{c_1, c_2, \cdots, c_n\}$ represent the CHNs distributed in the area. The UAV should select which CHNs to visit, and a sequence in which to visit them to simultaneously meet these two objectives.

The vectors of the CHN data sequence and amount of data collected are expressed as $\rho = [\rho_1, \rho_2, \cdots, \rho_k]^T$ and $D = [d^1, d^2, \cdots, d^k]^T$, respectively, where ρ_j represents the j_{th} visited CHN, d^j is the amount of data contained in the j_{th} CHN, and $k < n$. For each visiting target of the UAV, there is the constraint:

$$\rho_i \neq \rho_j, \; if \; i \neq j, \tag{1}$$

which means that a CHN can only be visited once.

The first goal of the UAV is to maximize the amount of data collected. For convenience, the objective function can be written as a minimization problem,

$$\min_{D} \; f_1 = -\sum_{i=1}^{k} d^i, \tag{2}$$

where d^i is the amount of data contained in the i_{th} CHN visited by the UAV, and k is the total number of UAVs accessing CHNs. It can be seen that the objective function decreases monotonically and is not positive.

The second goal of the UAV is to minimize its flight time, including time spent collecting data and flying to CHNs,

$$\min_{\rho, D} \; f_2 = t^{col} + t^{tra}, \tag{3}$$

where t^{col} is the hovering time of the UAV,

$$t^{col} = \sum_{i=1}^{k} \frac{d^i}{\epsilon}, \tag{4}$$

where d^i is the amount of data stored in the i_{th} CHN, ϵ is the data transmission rate between the UAV and the CHNs, and t^{tra} is the traveling time of the UAV:

$$t^{tra} = \sum_{i=1}^{k} \frac{dist(\rho(i), \rho(i+1))}{v}, \tag{5}$$

where $dist(\rho(i), \rho(i+1))$ is the Euclidean distance between $\rho(i)$ and $\rho(i+1)$, and v is the flight speed of the UAV.

The aim of global path planning for the multi-objective UAV-assisted SDC task is to determine which CHNs to visit, and the sequence in which to visit them. In this process, the UAV data collection amount is maximized and the UAV flight time is minimized, which can be expressed as

$$\min_{\rho, D} \quad \boldsymbol{f} = [f_1, f_2]$$
$$s.t. \quad (1), (2), (3), (4), (5), \ \rho_1 \in \{1, \cdots, k\}. \tag{6}$$

The optimization problem is now defined. We next propose a method to solve it.

4   Optimization Algorithm

In different problem scenarios, such as CHN location and data volume changes, traditional multi-objective optimization algorithms must compute from scratch, which is infeasible in practice. Methods based on reinforcement learning do not have this problem. In addition, we use a decomposition strategy [10], which is a simple and effective method to design MOPs.

We propose an effective DRL-based framework that can be utilized for global path planning in the multi-objective UAV-assisted SDC problem. The weighted sum [10] approach is adopted to decompose the MOP into a series of scalar optimization subproblems, each modeled as a neural network. A neighborhood-based parameter transfer strategy [8] and actor-critic algorithm are utilized to train the network and solve each subproblem, which completes the solution.

4.1   General Framework

Adopting the idea of "divide and conquer," decomposition has achieved good results in MOPs [10]. A set of uniformly distributed weight vectors $\boldsymbol{\lambda}^0, \boldsymbol{\lambda}^1, \cdots,$ $\boldsymbol{\lambda}^M$ is defined, where $\boldsymbol{\lambda}^i = (\lambda_1^i, \lambda_2^i)$, and M is the number of subproblems. The original MOP is decomposed into $M + 1$ scalar optimization subproblems,

$$\min_{\rho, D} \quad \boldsymbol{f}^i = \lambda_1^i f_1 + \lambda_2^i f_2$$
$$s.t. \quad (1), (2), (3), (4), (5), \ i \in \{0, 1, \cdots, M\}, \tag{7}$$

where $\lambda_1^i = 1 - \frac{1}{M}i$, $i = 0, 1, \cdots, M$, and $\lambda_2^i = 1 - \lambda_1^i$. The desired Pareto frontier can be obtained after these are solved.

It can be seen from (7) that two neighboring scalar optimization subproblems can have similar optimal solutions [10] because their weight vectors are numerically close. Thus, the neighborhood-based parameter transfer strategy can be utilized to collaboratively solve the $M + 1$ subproblems, accelerate network training, and obtain better solutions. We model each scalar optimization subproblem as a neural network, where the parameters of the i_{th} neural network are expressed as $\boldsymbol{P} = [\boldsymbol{w}_{\lambda^i}, \boldsymbol{b}_{\lambda^i}]$. The overall DRL framework is shown as Algorithm 1.

Algorithm 1. General framework of DRL

Input: Parameters of subproblem model $\boldsymbol{P} = [\boldsymbol{w}_{\lambda^i}, \boldsymbol{b}_{\lambda^i}], i = 0, 1, \cdots, M$; $M+1$ weight
    vectors $\boldsymbol{\lambda}^0, \boldsymbol{\lambda}^1, \cdots, \boldsymbol{\lambda}^M$
Output: Optimal parameters of subproblem model $\boldsymbol{P} = [\boldsymbol{w}^*, \boldsymbol{b}^*]$
  1: $[\boldsymbol{w}_{\lambda^0}, \boldsymbol{b}_{\lambda^0}] \leftarrow Random\_Initialize$
  2: **for** $i \leftarrow 0 : M + 1$ **do**
  3:     **if** $i == 0$ **then**
  4:         $[\boldsymbol{w}_{\lambda^0}^*, \boldsymbol{b}_{\lambda^0}^*] \leftarrow subproblem\_solving([\boldsymbol{w}_{\lambda^0}, \boldsymbol{b}_{\lambda^0}])$
  5:     **else**
  6:         $[\boldsymbol{w}_{\lambda^i}, \boldsymbol{b}_{\lambda^i}] \leftarrow [\boldsymbol{w}_{\lambda^{i-1}}^*, \boldsymbol{b}_{\lambda^{i-1}}^*]$
  7:         $[\boldsymbol{w}_{\lambda^i}^*, \boldsymbol{b}_{\lambda^i}^*] \leftarrow subproblem\_solving([\boldsymbol{w}_{\lambda^i}, \boldsymbol{b}_{\lambda^i}])$
  8:     **end if**
  9: **end for**
10: Return $[\boldsymbol{w}^*, \boldsymbol{b}^*]$

The inputs of Algorithm 1 are the initial parameters of each subproblem network $\boldsymbol{P} = [\boldsymbol{w}_{\lambda^i}, \boldsymbol{b}_{\lambda^i}]$ and its weight vector $\boldsymbol{\lambda}^i$, and the outputs are the optimized parameters of each subproblem $[\boldsymbol{w}_{\lambda^i}^*, \boldsymbol{b}_{\lambda^i}^*]$. Subproblems are solved on lines 4 and 7, as described in Sect. 4.2. We can use the optimized parameters of the $(i-1)_{th}$ subproblem network as the starting point of the i_{th} subproblem on line 6, and so on. This parameter transfer strategy makes use of the information between neighborhoods and solves the MOP based on deep reinforcement learning, which saves much calculation time. Other advantages of DRL are modularity and ease of use. For example, we can integrate other subproblem solvers into DRL, such as transformer [11], for global path planning. As shown in Algorithm 1, DRL is an outer loop, and the next step is to solve each scalar optimization subproblem obtained by decomposition.

4.2   Subproblem Solution

A modified pointer network [13] is used to solve the subproblems, and an actor-critic algorithm [12] is used to train the pointer network.

We introduce the input of the subproblem neural network, which is the status data of CHNs, $\boldsymbol{X} = \{\boldsymbol{x^1}, \cdots, \boldsymbol{x^n}\}$, where $\boldsymbol{x^i} = (a^i, b^i, d^i)$ is a tuple composed of

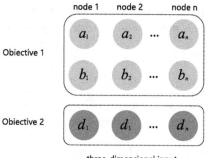

Fig. 2. Input form of subproblem neural network to solve global path planning for the SDC problem.

the geographic abscissa, ordinate, and data amount of the i_{th} CHN. The input is shown in Fig. 2. It is noteworthy that the data amount may change dynamically during processing. For instance, the amount of data stored in a node becomes zero when a UAV visits it, while location information remains unchanged. We use x_t^i to represent the state of input sample i at time t, and X_t is the set of all input states.

The output of the subproblem model is the permutation of a series of CHNs, $Y = \{\rho_t, t = 1, 2, \cdots, k\}$, where k is the number of CHNs to be accessed by the UAV, and ρ_t points to a CHN to determine the target of the next visit. The solution process ends when the length of the output sequence meets the demand. Y is calculated by decomposing the sequence using the chain rule,

$$P(Y|X) = \prod_{t=1}^{k} P(\rho_{t+1}|\rho_1, \cdots, \rho_t, X_t). \tag{8}$$

An arbitrary CHN is chosen as ρ_1. We choose ρ_{t+1} from the CHNs that have not been visited by the UAV. Once the UAV visits a CHN, X_t is updated. In short, according to the visited CHNs and the X_t information of the current step, we use (8) to determine which CHN to visit next. Pointer network [13] just has a right ability to remember the previous information.

A modified pointer neural network is utilized to model (8). It is a sequence-to-sequence model that maps one sequence to another. The model consists of an encoder and decoder, which are both recurrent neural networks (RNNs). The encoder maps the input low-dimensional vector x^j to the high-dimensional vector e^j, which contains highly abstract information. However, since the change of CHN order in the input should not affect the outcome, we use one-dimensional convolution instead of RNN to reduce the computational complexity without diminishing performance. The decoder obtains the visit sequence of CHNs according to the high-dimensional information generated by the encoder. With the output of the encoder and the hidden state h_t of the RNN decoder, which includes information on previously visited CHNs, the conditional probability of accessing the next node can be calculated using an attention mechanism,

$$\boldsymbol{u}_j^t = W_1^T tanh(W_2 \boldsymbol{e}^j + W_3 \boldsymbol{h}_t), j \in (1, \cdots, n)$$
$$P(\rho_{t+1}|\rho_1, \cdots, \rho_t, X_t) = softmax(\boldsymbol{u}^t)$$

$$(9)$$

where W_1, W_2, and W_3 are trainable parameters, and $\boldsymbol{u}_j^t$ is the probability of visiting the j_{th} CHN in step t. Based on a greedy strategy, we take the CHN with the highest probability as the next CHN for the UAV to visit in the prediction stage. However, to search for potentially better solutions, we instead use a sampling method to select the next node to visit.

Algorithm 2. Actor-critic training algorithm
___

Input: $\theta, \phi \leftarrow$ initialized parameters given in Algorithm 1
Output: Optimal parameters θ, ϕ
1: **for** $iteration \leftarrow 1,2,\cdots$ **do**
2:     reset gradients: $d\theta \leftarrow 0, d\phi \leftarrow 0$
3:     generate n problem samples from $\{\Phi_{M_a}, \Phi_{M_b}, \Phi_{M_d}\}$
4:     **for** $i \leftarrow 1, \cdots,$ n **do**
5:         $t \leftarrow 0$
6:         **while** not terminated **do**
7:             select the next CHN ρ_{t+1}^i
                according to $P(\rho_{t+1}^i|\rho_1^i, \cdots, \rho_t^i, X_t^i)$
8:             Update X_t^i to X_{t+1}^i by leaving out visited
                CHNs
9:         **end while**
10:        Compute the reward R^i
11:    **end for**
12:    $d\theta \leftarrow \frac{1}{n} \sum\limits_{i=1}^{i=n} (R^i - V(X_0^i; \phi)) \nabla_\theta logP(Y^i|X_0^i)$
13:    $d\phi \leftarrow \frac{1}{n} \sum\limits_{i=1}^{i=n} \nabla_\phi (R^i - V(X_0^i; \phi))^2$
14:    $\theta \leftarrow \theta + \eta d\theta$
15:    $\phi \leftarrow \phi + \eta d\theta$
16: **end for**

We use the actor-critic algorithm, shown as Algorithm 2, to train the network: (i) the actor network, which is the modified pointer neural network, provides the probability of visiting the next CHN; and (ii) the critic network estimates the reward according to the problem state.

The inputs of Algorithm 2 are θ and ϕ, which are the parameters of the actor and critic network, respectively. We point out that the actor network is the previous pointer network and $\theta = \boldsymbol{P}$. The outputs are the optimized θ and ϕ. We train the network in an unsupervised way and generate training data $\boldsymbol{X}$ from normal distribution $\{\Phi_{M_a}, \Phi_{M_b}, \Phi_{M_d}\}$, where Φ_{M_a}, Φ_{M_b}, and Φ_{M_d} are respectively the geographic abscissa and ordinate, and the data amount of CHN.

We sample n problem samples from $\{\Phi_{M_a}, \Phi_{M_b}, \Phi_{M_d}\}$, and each is utilized by the actor network to generate a visiting sequence of CHNs with current parameters θ. According to this sequence and objective function (7), the reward R^i of the i_{th} sample is calculated at line 10. The parameters of the actor network θ are updated using the policy gradient calculated at line 14. Here, $V(X_0^i; \phi)$ is the reward of the i_{th} sample estimated by the critic network. By reducing the gap between the estimated and real rewards, the parameters of the critic network ϕ can be updated at line 15, so as to optimize θ and ϕ together.

When the $M + 1$ subproblems are solved, the Pareto optimal solution of the original MOP can be obtained directly through the forward propagation of the models. Unlike traditional multi-objective evolutionary algorithms, our algorithm does not have to compute from scratch, and is more robust.

5   Performance Evaluation and Analysis

We compare the solutions obtained by our method and NSGA-II [15] and MOEA/D [10], which are two classical multi-objective evolutionary algorithms. The genetic algorithm NSGA-II reduces the complexity of non-inferior sorting genetic algorithms, runs quickly, and converges well. MOEAD decomposes a MOP into several scalar optimization subproblems, and optimizes them simultaneously through the information of several adjacent subproblems.

5.1   Setting and Parameters

A UAV flies in a square area of 100 km^2, in which CHNs are randomly distributed. We train and test in the cases of 40 and 50 CHNs. Based on experience, we suppose that collecting half of the header nodes in the area will meet the task requirements. The amount of data to be collected from each CHN is randomly distributed between 0 and 20 GB. The flight speed and data collection rate of the UAV are 30 km/h and 12 MB/s, respectively. The number of iterations of NSGA-II and MOEA/D are set to 500, 1000, 2000, and 4000, and the population size is 100. We set the number of subproblems of DRL to 100. We use 120000 randomly generated global path planning instances to train our DRL framework. We choose as performance indicators the hypervolume (HV) [14], which is a comprehensive index to evaluate the convergence and diversity of solutions, and execution time, as shown in Table 1.

5.2   Results and Discussion

We compare the Pareto front obtained by our method and traditional multi-objective evolutionary algorithms with 50 CHNs, as shown in Fig. 3. Our method can obviously collect more data with various flight times. The flight time of the UAV with our DRL method is the least with various amounts of data collection. The Pareto front obtained by our method is closer to the true Pareto front. In addition, combined with the HV values in Table 1, our method can be found to

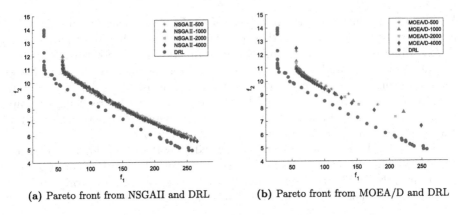

(a) Pareto front from NSGAII and DRL   (b) Pareto front from MOEA/D and DRL

Fig. 3. Randomly generated global path planning for data collection problem of 50 CHNs: PF comparison of NSGA-II, MOEA/D, and DRL method (trained on 50 nodes).

have better convergence and diversity than the two traditional multi-objective evolutionary algorithms. In the case of 50 nodes, NSGA-II can improve performance by increasing the number of iterations when it focuses more on the time spent to complete the task. However, the gain is trivial when balanced against its increased computational complexity. Similarly, when more attention is paid to the amount of data collection, MOEA/D can achieve a slight improvement by increasing the number of iterations. However, it cannot yield a satisfactory Pareto front when it focuses more on the time to complete the task. In a nutshell, our DRL method can balance two conflicting objective functions and provide a Pareto front with satisfactory convergence and diversity.

We compare the Pareto front obtained by our method and traditional multi-objective evolutionary algorithms with 40 CHNs, as shown in Fig. 4. Similar to the case of 50 CHNs, our DRL method has better convergence and diversity, and its Pareto front is closer to the true Pareto front. We emphasize another advantage of the DRL method. As previously stated, instead of an RNN, the encoder part of the actor network utilizes one-dimensional convolution to map the input to high-dimensional space. Because the parameters of the one-dimensional convolution layer are shared among all CHNs, the encoder is robust to the change of CHNs. Therefore, when the number of CHNs changes, the method does not need to be retrained, and the Pareto front in the new circumstance can be obtained directly through forward propagation.

The experimental results show the effectiveness of our method in global path planning for the multi-objective UAV-assisted SDC problem.

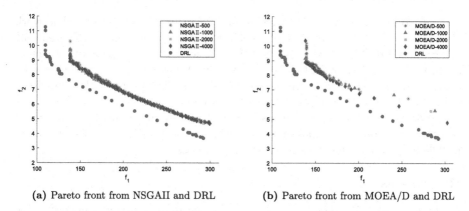

(a) Pareto front from NSGAII and DRL   (b) Pareto front from MOEA/D and DRL

Fig. 4. Randomly generated global path planning for data collection problem of 40 CHNs: PF comparison of NSGA-II, MOEADm, and DRL method (trained on 50 nodes).

We compare the execution time of the algorithms with 40 and 50 CHNs and 500, 1000, 2000, and 4000 iterations, as shown in Table 1. Once training is completed, our DRL method has excellent advantages in running time.

Table 1. Comparison of algorithm execution time and HV values

	40 CHNS		50 CHNS	
	Time/s	HV	Time/s	HV
NSGAII-500	4.75	1104	4.85	1686
NSGAII-1000	9.46	1186	10.92	1722
NSGAII-2000	18.39	1123	21.15	1714
NSGAII-4000	40.00	1117	41.67	1722
MOEA/D-500	21.91	953	22.68	1453
MOEA/D-1000	45.50	990	45.22	1514
MOEA/D-2000	91.67	1030	94.29	1566
MOEA/D-4000	186.80	1062	189.83	1591
DRL	**1.91**	1400	**2.30**	2053

6   Conclusions

We presented a DRL method to solve the global path planning for a multi-objective UAV-assisted SDC problem. By selecting the CHNs and controlling the visiting order, the amount of data collection was maximized, and the UAV flight time was minimized. We decomposed the problem into a series of scalar optimization subproblems, and used a neighborhood-based parameter transfer strategy to accelerate and optimize training. We utilized a pointer neural network and an actor-critic algorithm to solve each subproblem. Experiments showed that the proposed method is effective, and is superior to traditional multi-objective evolutionary algorithms NSGA-II and MOEA/D in convergence, diversity, time complexity, and robustness. In view of our results, our proposed UAV-assisted data collection method has wide potential application in WSNs.

MOP based on DRL is still in its infancy. In the future, we will consider more complex situations in global path planning for multi-objective UAV-assisted SDC problems, such as three or four objective functions.

References

1. Li, C., Zhang, H., Hao, B., Li, J.: A survey on routing protocols for large-scale wireless sensor networks. Sensors (Basel, Switzerland)**11**(4) (2011)
2. Li, B., Fei, Z., Zhang, Y.: UAV communications for 5g and beyond: Recent advances and future trends. IEEE Internet Things J. **6**(2), 2241–2263 (2019)
3. Zhan, C., Zeng, Y., Zhang, R.: Trajectory design for distributed estimation in UAV-enabled wireless sensor network. IEEE Trans. Veh. Technol. **67**, 10155–10159 (2018)
4. Wang, Y., et al.: Trajectory design for UAV-based internet-of-things data collection: a deep reinforcement learning approach. IEEE Internet of Things J. pp. 1–1 (2021)
5. Gong, J., Chang, T.H., Shen, C., Chen, X.: Flight time minimization of UAV for data collection over wireless sensor networks. IEEE J. Select. Areas Commun. **36**, 1942–1954 (2018)
6. Zeng, Y., Zhang, R.: Energy-efficient uav communication with trajectory optimization. IEEE Trans. Wireless Commun. **16**(6), 3747–3760 (2017)
7. Ghdiri, O., Jaafar, W., Alfattani, S., Abderrazak, J.B., Yanikomeroglu, J.B.: Offline and online UAV-enabled data collection in time-constrained IoT networks. IEEE Trans. Green Commun. Netw. **5**, 1918–1933 (2021)
8. Li, K., Zhang, T., Wang, R.: Deep reinforcement learning for multiobjective optimization. IEEE Trans. Cybernet. **51**(6), 3103–3114 (2021)
9. Shu, F., et al.: UAV path intelligent planning in IoT data collection. J. Commun. **42**(2), 10 (2021)
10. Zhang, Q., Li, H.: Moea/d: a multiobjective evolutionary algorithm based on decomposition. IEEE Trans. Evol. Comput. **11**(6), 712–731 (2007)
11. Xu, Y., Fang, M., Chen, L., Xu, G., Du, Y., Zhang, C.: Reinforcement learning with multiple relational attention for solving vehicle routing problems. IEEE Trans. Cybernet. **52**, 11107–11120 (2021)
12. Mnih, V., et al.: Asynchronous methods for deep reinforcement learning. In: International Conference on Machine Learning, pp. 1928–1937 (2016)
13. Nazari, M., Oroojlooy, A., Takáč, M., Snyder, L.V.: Reinforcement learning for solving the vehicle routing problem. In: NIPS 2018 Proceedings of the 32nd International Conference on Neural Information Processing Systems, vol. 31, pp. 9861–9871 (2018)
14. Shang, K., Ishibuchi, H., He, L., Pang, L.M.: A survey on the hypervolume indicator in evolutionary multiobjective optimization. IEEE Trans. Evol. Comput. **25**(1), 1–20 (2021)
15. Deb, K., Pratap, A., Agarwal, S., Meyarivan, T.: A fast and elitist multiobjective genetic algorithm: NSGA-II. IEEE Trans. Evol. Comput. **6**(2), 182–197 (2002)

Infrared Image Object Detection of Vehicle and Person Based on Improved YOLOv5

Jintao Wang[1], Qingzeng Song[2], Maorui Hou[2], and Guanghao Jin[3](✉)

[1] School of Software, Tiangong University, Tianjin 300387, China
[2] School of Computer Science and Technology, Tiangong University, Tianjin 300387, China
[3] Beijing Polytechnic, Beijing 100176, China
jingh_research@163.com

Abstract. Existing object detection algorithms are difficult to perform object detection tasks on embedded devices under the limitations of energy efficiency ratio and power consumption due to complex network structure and huge computational and parametric quantities. The object detection task in infrared images has low recognition rate and high false alarm rate due to long distance, weak energy and low resolution. In order to achieve the detection task at the mobile edge of infrared vehicle pedestrian target detection, this paper puts the YOLOv5 algorithm into a series of optimizations and proposes a lightweight YOLO-mini network structure. That is, instead of CSPDarknet, the MobileNetV2 network structure is used as the backbone feature extraction network with the addition of coordinate attention mechanism. Also, to make the network model more lightweight, the weights are converted to int8 type by quantized sensing training, which enables the task of the object detection algorithm for infrared vehicle pedestrian dataset on embedded devices. Experiments testing the FLIR dataset on NVIDIA Xavier NX show that this algorithm greatly reduces the number of network model parameters with less loss of accuracy and improves the FPS. mAP of YOLO-MobileNetV2 reaches 86.75%, number of parameters 2.76M, and FPS of 45; The network structure of YOLO-mini achieves 84.63% mAP, 0.69M number of parameters, and 63 FPS.

Keywords: Object Detection · YOLOv5 · Quantization · Coordinate Attention · NVIDIA Xavier NX

1 Introduction

Target detection in infrared images has received a lot of attention in the field of computer vision because of its important application research value. It also occupies an irreplaceable position in many fields such as lesion cell diagnosis [1], video surveillance [2], UAV cruise [3], infrared early warning [4], infrared night vision [5], infrared guidance [6], and other civil and military fields. Compared with visible images, infrared images have longer imaging wavelengths, greater noise, poor spatial resolution, and are sensitive to changes in ambient temperature, so traditional target detection methods in infrared images are limited by a single method.

S. Yang and S. Islam (Eds.): APWeb-WAIM 2022 Workshops, CCIS 1784, pp. 175–187, 2023.
https://doi.org/10.1007/978-981-99-1354-1_16

Convolutional neural networks can learn high-level data features from the raw pixels of training data to obtain better feature representation of complex contextual information. The current deep learning algorithms can be broadly classified into two-stage algorithms and one-stage algorithms. Among the two-stage target detection algorithms include Rcnn [7], Fast-Rcnn [8], Faster-Rcnn [9], etc. These algorithms can obtain high accuracy, but the training steps are tedious and occupy considerable space. One-stage target detection algorithms, such as SSD [10] and YOLO [11] algorithms do not require candidate frames and perform classification and regression prediction directly by convolution operations. This class of algorithms is fast and has a small space footprint. Due to the need of low power consumption and low energy consumption on embedded platforms.

In this paper, the network structure is studied based on the YOLOv5 [15] algorithm, and the lightweight MobileNetv2 [16] design is used to make it meet the requirements of micro-arithmetic power and low power consumption in the FLIR infrared vehicle pedestrian target detection dataset task. A new detection model, YOLO-mini, is proposed.

2   Related Work

Infrared image object detection has the advantage of not being disturbed by the environment, and it is a research hotspot in the field of object detection [17]. Currently, infrared image object detection approaches have been divided into two types: traditional algorithms and CNN models.

Traditional infrared object detection methods consider infrared images as three parts: object, background, and noise in the images. The idea is to suppress background and noise, thus, strengthening the object to achieve object detection by using various methods. Zhao and Kong [18] first used the detection method based on spatial filtering for infrared object detection. In terms of different gray values of object and background, the background is selected and suppressed, and thus the object is detected. However, this method allowed all isolated noise points of small objects to pass, leading to a low detection rate. To address this problem, Anju and Raj [19] used the frequency difference between the object and background to separate the high-frequency part and the low-frequency part to achieve the detection task. Compared with that of spatial filtering methods, the detection effect of frequency domain filtering is a substantial promotion, but incurs high computational complexity. Jiao [20] adopted sparse representation to cast the principle of infrared object detection in the form of a low-rank matrix and sparse matrix recovery, thus, achieving object segmentation and detection. Unfortunately, these infrared object detection algorithms have strong pertinence, a high time complexity, and window redundancy. They are also not robust to the changes in object diversity.

Due to the improvement in computability and the widespread use of infrared imaging system equipment, many datasets are released to the public, such as KAIST [21], FLIR [22], and OTCBVS [23], which prompts deep learning to be gradually applied in the field of infrared image object detection. Thanks to the strong capability of feature expression, the CNN models open new horizons and create a large amount of excitement in object detection. They preserve the neighborhood relations and spatial locality of the input in their latent higher-level feature representations. Additionally, the number of free parameters describing their shared weights does not depend on the input dimensionality, meaning that the CNN can scale well to realistic-sized high-dimensional images in terms of computational complexity.

M. Li et al. proposed SE-YOLO [24], a real-time pedestrian object detection algorithm for small objects in infrared images, which improved the feature expression ability of the network combined with the SE block.To further improve the speed and accuracy of object detection, especially when objects are small and occluded, Li et al. [25] developed a detector, YOLO-ACN, by introducing an attention module, CIoU (Complete Intersection over Union) loss, improved Soft-NMS (Non-Maximum Suppression), and depthwise separable convolution. The detector, YOLOACN, can focus on small objects and avoid the deletion of occluded objects. However, there were still a large number of parameters to save that the weight file was too large, which made it difficult to apply on mobile devices. In addition to the above methods based on the classic YOLOv3, there were some other one-stage network models for infrared image object detection. Cao et al. [26] presented a DNNbased one-stage detector, ThermalDet, which included a dual-pass fusion block (DFB) and a channelwise enhancement module (CEM). The mAP of ThermalDet is 74.6% in the FLIR dataset, and thus, cannot achieve the desired results. Song et al. [27] harnessed the features of infrared images and visible images to achieve fused features. Then, a multispectral feature fusion network (MSFFN) was proposed based on YOLOv3 to detect pedestrian objects, but the excellence of the MSFFN was only obvious when the input images were of a small size.

These methods achieved a better performance for nighttime object detection in different fields, such as pedestrian detection and autonomous driving. Despite the recent progress, it was difficult to transplant these models to mobile devices after training, especially for drone equipment, satellite equipment, infrared cameras, etc.

To address the problems of existing models, this paper investigates the state-of-the-art target detection algorithm YOLOv5, a detector first released on 25 June 2020. Based on the study of the unique features of infrared images, this paper proposes YOLO-mini a vehicle pedestrian infrared target method with YOLOv5 as the core, by optimising the network structure, compressing channels, quantization, optimising parameters, and incorporating an improved coordinate attention module in the residual block to improve feature extraction and adding a detection layer to detect smaller objects. Experimental results show that the proposed infrared image target detection model YOLO-mini has significant improvements in detection accuracy, detection speed and model size compared to the latest infrared image target models on the FLIR dataset.

3   Improvement of YOLOv5

3.1   Improvement of Backbone Network

MobileNet16series networks, as representatives of lightweight deep convolutional neural networks, are widely used in embedded and mobile terminals. In this paper, the YOLOv5s Backbone backbone network is replaced by MobileNetv2 for feature extraction with fewer parameters, faster speed, and lower memory consumption, and the features are extracted by deep separable convolution instead of the original convolutional layer, which increases the computational speed while reducing the number of parameters, and also significantly reduces the demand for computing power. The YOLO-MobileNetV2 feature extraction network structure is shown in Table 1.

Table 1. YOLO-MobileNetV2 network structure

Original Input	Ours Input	Operator		T	C	N	S
$224^2 * 3$	$640^2 * 3$	**Conv2d**	√	-	32	1	2
$112^2 * 32$	$320^2 * 32$	**bottleneck**	√	1	16	1	1
$112^2 * 16$	$320^2 * 16$	**bottleneck**	√	6	24	2	2
$56^2 * 24$	$160^2 * 24$	**bottleneck**	√	6	32	3	2
$28^2 * 32$	$80^2 * 32$(**FPN**)	**bottleneck**	√	6	64	4	2
$14^2 * 64$	$40^2 * 64$	**bottleneck**	√	6	96	3	1
$14^2 * 96$	$40^2 * 96$(**FPN**)	**bottleneck**	√	6	160	3	2
$7^2 * 160$	$20^2 * 160$	**bottleneck**	√	6	320	1	1
$7^2 * 320$	$20^2 * 320$(**FPN**)	Conv2d 1*1	–	–	1280	1	1
$7^2 * 1280$	–	Avgpool7*7	–	–	–	1	–
$1 * 1 * 1280$	–	Conv2d1*1	–	–	k	–	–

In the detection layer, three branches of different scales are made to be output and extracted at Bottleneck's layers 7, 13 and 17 respectively, discarding the modules after layer 17; each of these three layers is then transformed by a PW convolution operation to transform its dimensionality so that it can be connected to the detection layer.

3.2 Addition of Coordinate Attention

In recent years, attention mechanism modules have been widely used in computer vision tasks. However, the use of attention mechanisms in lightweight networks is somewhat limited because the additional computational overhead associated with most attention mechanisms is not affordable for lightweight networks. In this paper, we introduce a simple and flexible Coordinate Attention [28] (CA) that imposes little additional computational overhead to improve the accuracy of the network. The flow of the CA module is shown in Fig. 1.

As shown in the figure above (Coordinate Attention), two 1D vectors are obtained by average pooling in the horizontal X direction and the vertical Y direction respectively, and the channels are compressed by Concat and 1x1Conv in the spatial dimension, and then encoded by BN and Non-linear. The spatial information in the vertical and horizontal directions is then split, and then each passes through 1x1Conv to obtain the same number of channels as the input feature maps, and finally performs normalization and weighting. To put it simply, Coordinate Attention performs average pooling in the horizontal and vertical directions, then encodes the spatial information, and finally fuses the spatial information by weighting on the channel.

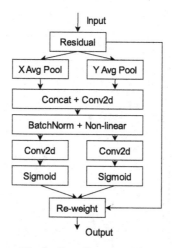

Fig. 1. Coordinate attention

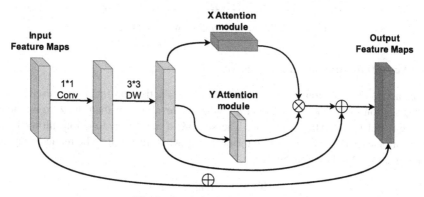

Fig. 2. Bottleneck + CA module

So far it is able to effectively focus on the effective channels while paying attention to the spatial location coordinate information. As shown in Fig. 2, this paper embeds this attention mechanism into the new Backbone, the inverted residual structure of the linear bottleneck in MobileNetv2, to help the model be more capable of multi-dimensional feature extraction for targets of interest. For example, more attention is paid to vehicle and pedestrian targets with small targets, greatly enhancing the efficiency of model training and the accuracy of recognition. The improved YOLOv5 framework in this paper is shown in the following Fig. 3.

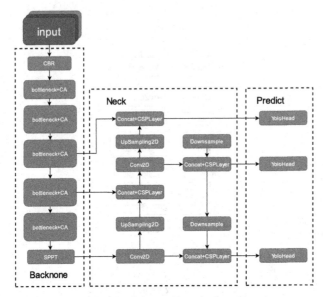

Fig. 3. YOLO-mini overall network architecture

3.3 Activation Function Improvement

At the activation function level, the Mish non-linear activation function is used, which adds a section of smoothed points at the negative level, rather than the zero boundary as in the ReLU activation function. The smoothed activation function allows better information to penetrate deeper into the neural network, resulting in better accuracy and generalisation. As shown in Eq. (1).

$$Mish = x \times \tanh(1 + e^x) \tag{1}$$

where x is the value of the parameter passed in via the normalisation layer.

3.4 Quantization and Training of Neural Networks

We chose Pytorch version 1.40 as the framework for 8bit quantization. PyTorch offers three methods for quantizing models: dynamic quantization after training, static quantization after training, and perceptual quantization training. This is usually done when memory bandwidth and computational savings are important, and CNNs are the typical use case [29].

In this paper, a post-training 8bit static quantization training method is used to reduce the number of parameters by changing our model weights to the INT8 type and compressing our model while maximising accuracy retention.

Before the quantization training, there is another important step which is to perform model fusion, i.e. to fuse the three layers of convolutional, batch normalisation and activation function layers before the subsequent quantization operation can be performed. The activation functions in MobileNetv2 use h-Swish and ReLU6, which prevented fusion during perceptual quantization; therefore, they were both changed to ReLU activation functions before the model fusion operation could be performed.

4 Experiment

4.1 Experimental Environment

The experiments in this paper use the Pytorch framework for network structure modification and quantization operations, training on four GPUs on a docker-deployed cluster, and inference and testing operations on an embedded device, the Nvidia Jetson NX.

4.2 FLIR Dataset

We use FLIR ADAS22as the dataset, which consists of 9214 thermal images annotated with borders. Each image has a resolution of 640×512 and was captured by a FLIR-Tau2 camera. 60% of the images were taken during the day and 40% of the images were taken at night. This dataset provides visible spectrum (RGB) images and infrared thermal images. In our experiments, only infrared thermal images of the dataset are considered. We conduct experiments using the training and test splits suggested in the dataset documentation. The objects in the dataset are divided into four categories, namely bicycles, cars, dogs, and people. However, the dog class has very few annotations, only 200 samples, and uses teaching tricks in the actual human-car scene. Therefore, dogs were not considered in our experimental study. We crop the square image with the bounding box annotation of the object. After extracting the objects, we resize the image to 640×640.

Finally, after filtering our dataset consists of 4457 bicycle labels, 46692 car labels and 28151 person image labels. The image below shows an example image from the FLIR ADAS dataset (Fig. 4).

Fig. 4. FLIR original image

4.3 Experimental Equipment

The embedded device uses the Jetson Xavier NX, the latest GPU edge computing device from Nvidia in 2020, which delivers up to 14 terra operations per second (TOPS) at 10W or 21 terra operations per second at 15W, running multiple neural networks in parallel. Linux operating system.

We installed the Ubuntu 18.04 operating system image on the embedded device and configured the Python 3.6 and PyTorch environments for subsequent testing experiments on the device (Fig. 5).

Fig. 5. Jetson Xavier NX

4.4 Experimental Settings

In this paper, the size of the input image is set to 640 * 640 jpeg images, and the training is performed on four GPUs on the docker-deployed cluster; the CIOU Loss is used as the loss function, the total Batch size is set to 16, the Batch size of each GPU is 4, the Epoch is set to 600, and the learning rate is set to 0.0001, simulating the cosine annealing strategy to adjust the learning rate of the network. The mosaic data augmentation method was used for data augmentation, and the intra-group label smoothing method was used to optimize the label classes. As the original anchor size was too large compared to the target of the human-vehicle image, this thesis used the K-means clustering algorithm to select the most appropriate nine prior frames for the input dataset (10,23),(15,18),(14,37),(26,28),(20,63),(43,43),(37,117),(76,72),(131,150) and applied them to the two networks. The original YOLOv5s network and the improved

YOLOv5-MobileNetv2 network, and the YOLO-mini network trained by perceptual quantization were trained respectively, and finally the results of the above three models were compared (Fig. 6).

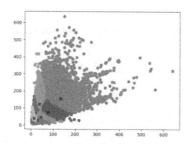

Fig. 6. K-means result

4.5   Results and Analysis

When measuring the performance of a model, Prescision and Recall are often used to evaluate the accuracy of its network predictions. However, in the field of target detection, the AP (Average Precision) metric is often used to measure the accuracy of a particular type of target detection for single-target detection algorithms. For multi-target detection algorithms, mAP (mean Average Precision) is usually used to measure the average precision of multi-target detection results. As this paper is a target detection task on a 3-classified infrared dataset, the mAP metric is used. The final inference detection is performed on an embedded device, i.e. to maximise model size, reduce model computations (FLOPs) and improve frame rates (FPS) on the embedded device without reducing mAP or reducing the limited mAP (Table 2).

Table 2. Ablation experiments on the test set of FLIR

Method	bicycle	Car	Person	mAP	Para	FPS
YOLOv5s	84.92	91.37	89.18	88.49	7.2M	17
YOLOv5 + MobileNetv2	80.86	87.21	84.89	84.32	2.53M	47
YOLOv5 + MobileNetv2 + CA(ours)	83.19	89.76	87.28	86.75	2.76M	45
YOLO-mini(ours)	82.41	88.56	85.92	85.63	0.69M	63

The number of model parameters for YOLOv5s was 46.5 MB, and the size of the model decreased to 2.53 MB after the feature extraction network was replaced with a depth-separable convolution.

The YOLOv5-MobileNetv2 method, with a maximum mAP of 86.75%, greatly compresses the model size and improves the detection speed, with a model parameter count of

2.7M The number of model parameters is 2.7M, which is only 38.3% of that of Yolov5s, and the mAP is only 1.74% points lower; the perceptually quantized model YOLO-mini is only 0.69M, which is 9.6% of that of Yolov5s, and the mAP is only 2.86% points lower.

Table 3. Accuracy comparison with advanced detectors

Method	bicycle	Car	Person	mAP	Para
Faster R-CNN30	54.7	67.6	39.6	53.9	-
MMTOD-CG30	63.3	70.6	50.3	61.4	-
RefineDet32	57.2	54.5	77.2	72.9	-
TermalDet31	60.0	85.5	78.2	74.6	-
YOLO-FIRI33	74.1	90.6	85.8	83.5	7.20M
YOLOv5 + MobileNetv2 + CA (ours)	**83.19**	**89.76**	**87.28**	**86.75**	**2.76M**
YOLO-mini(ours)	**82.41**	**88.56**	**85.92**	**85.63**	**0.69M**

As shown in Table 3, we compared the AP values for each category and the mAP of the different classification networks in percentage terms.

Our proposed methods YOLOv5-MobileNetv2, YOLO-mini are compared with the state-of-the-art methods in the table. In [30], the mAP was 53.9% when using only the faster R-CNN as a detector. However, MMTOD-CG uses a faster R-CNN as the backbone feature network and incorporates a pre-trained detector, resulting in a 7% increase in mAP. TermalDet [31] is based on RefineDet [32], which considers features from each layer as the final detection and increases the accuracy to 74.6%. Yolo-FIRI [33] improves on YOLOv5 with an improved detection head and the addition of multiscale detection, increasing the mAP to 83.5% (Fig. 7).

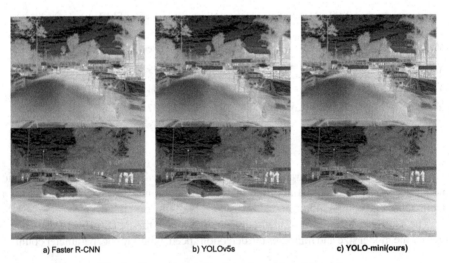

a) Faster R-CNN   b) YOLOv5s   c) YOLO-mini(ours)

Fig. 7. Comparison of improved experimental results

A comparison of the test images shows that YOLO-mini and YOLOv5s have similar test results, much higher than Faster R-CNN, but the model size and test speed of YOLO-mini are better than YOLOv5s.

The YOLOv5-Mobilenetv2, YOLO-mini model proposed in this paper is based on the state-of-the-art target detection network YOLOv5, which utilises the coordinate attention mechanism to avoid introducing large overheads by embedding the location information into the channel attention, thus allowing the mobile network to acquire information over a larger area, greatly reducing the number of network parameters and ensuring a higher mAP. We can observe that the mAP values are all better than in previous work, with our YOLOv5-Mobilenetv2 further reaching 86.75%, 3.2% higher than YOLO-FIRI; the number of model parameters is 2.7M, only 38.3% of YOLO-FIRI. The quantified network, YOLO-mini, had an mAP of 85.63, 2.1% higher than YOLO-FIRI; the number of model parameters was 0.69M, only 9.6% of YOLO-FIRI.

5   Conclusions

This paper designs a set of solutions for neural network acceleration algorithms in low-power application scenarios, and solves the contradiction between high computing requirements and low power consumption requirements such as RCNN series methods, SSD algorithms, and YOLO series algorithms in the embedded field. A lightweight neural network is designed for this task, and the network is quantitatively processed, which can ensure high-accuracy detection when the amount of model calculation and parameters is small. The neural network acceleration algorithm implemented this time can meet expectations in terms of performance, power consumption, and operating efficiency, and can recognize infrared images in low-power scenarios. In the next step, a larger-scale model compression method will be adopted, such as channel pruning algorithm or space pruning algorithm, and the combination of pruning algorithm and quantization algorithm will be used to further improve the scale and efficiency of model compression.

References

1. He, K., Zhang, X., Ren, S., Sun, J.: Deep residual learning for image recognition. In: Proceedings of IEEE Conference on Computer Vision and PatternRecognitio, pp. 770–778 (2016)
2. Zhu, H., Qin, L., Sun, B.: Review on parallelization of deep neural networks. J. Chin. J. Computer. 41(8), 171–191 (2018). https://doi.org/10.11897/SP.J.1016.2018.01861
3. Krizhevsky, I.S., Hinton, G.E.: ImageNet classification with deep convolutional neural networks. In: Proceedings of Advance Neural Information and Processing Systems, pp. 1097–1105 (2021)
4. Nair, V., Hinton, G.E.: Rectified linear units improve restricted Boltzmann machines. In: Proceedings of International Conference on Machine Learning, pp. 807–814 (2010)
5. Hinton, G.E., Srivastava, N., Krizhevsky, A., Sutskever, I., Salakhutdinov, R.R.: Improving neural networks by preventing coadaptation of feature detectors. arXiv:1207.0580 (2012)
6. Girshick, R., Donahue, J., Darrell, T., Malik, J.: Rich feature hierarchies for accurate object detection and semantic segmentation. In: Proceedings of IEEE Conference on Computer Vision and Pattern Recognition, pp. 580–587 (2014)

7. Girshick, R., Donahue, J., Darrell, T., et al.: Rich feature hierarchies for accurate object detection and semantic segmentation. In: IEEE Conference on Computer Vision and Pattern Recognition, pp. 580–587. IEEE Computer Society (2014)

8. Girshick, R.: Fast R-CNN. In: Proceedings of the IEEE International Conference on Computer Vision, pp. 1440–1448 (2015)

9. Ren, S., He, K.,Girshick, R., et al.: Faster R-CNN: towards real-time object detection with region proposal networks. In: Proceedings of the 2015 Advances in Neural Information Processing Systems, pp. 91–99 (2015)

10. Liu, W., et al.: SSD: single shot multibox detector. In: Leibe, B., Matas, J., Sebe, N., Welling, M. (eds.) Computer Vision – ECCV 2016. LNCS, vol. 9905, pp. 21–37. Springer, Cham (2016). https://doi.org/10.1007/978-3-319-46448-0_2

11. Redmon, J.,Divvala, S., Girshick, R., et al.: You Only look once: unified, real time object detection. In: Computer Vision and Pattern Recognition, pp. 6517–6525 (2017)

12. Redmon, J., Farhadi, A.: YOLO 9000: better, faster, stronger. In: IEEE Conference on Computer Vision and Pattern Recognition, pp. 6517–6525 (2017)

13. Redmon, J., Farhadi, A.: Yolov3: an incremental improvement. In: IEEE Conference on Computer Vision and Pattern Recognition.arXiv:1804.0276 (2018)

14. Bochkovskiy, A., Wang, C.Y., Liao, H.Y.M.: YOLOv4: optimal speed and accuracy of object detection. In: IEEE Conference on Computer Vision and Pattern Recognition. arXiv:2004. 10934v1 (2020)

15. Jocher, G., Stoken, A., Borovec, J.: Ultralytics/yolov5: V4. 0-Nn. SiLU () activations weights & biases logging PyTorch hub integration. Zenodo, Techical report. https://zenodo. org/record/4418161. https://doi.org/10.5281/zenodo.4418161(2021)

16. Sandler, M., Howard, A., Zhu, M., et al.: MobileNetV2: inverted residuals and linear bottle-necks. In: Proceedings of the IEEE Conference on Computer Vision and Pattern Recognition, pp. 4510–4520 (2018)

17. Fang, L., Wang, X., Wan, Y.: Adaptable active contour model with applications to infrared ship target segmentation. J. Elect. Imaging 25(4), 1–10 (2016). https://doi.org/10.1117/1.JEI. 25.4.041010

18. Zhao, K., Kong, X.: Background noise suppression in small targets infrared images and its method discussion. Opt. Optoelectron. Technol. 2, 9–12 (2004)

19. Anju, T.S., Raj, N.R.N.: Shearlet transform based image denoisingusing histogram thresh-olding. In: Proceedings of International Conference on Communication System Network (ComNet), July 2016, pp. 162–166 (2016)

20. Jiao, P.: Research on image classification and retrieval method based on deep learning and sparse representation. M.S. thesis, Xi'an University Technology, Xi'an, China (2019)

21. Choi, Y., et al.: KAIST multi-spectral day/night data set for autonomous and assisted driving. IEEE Trans. Intell. Transp. Syst. 19(3), 934–948 (2018). https://doi.org/10.1109/TITS.2018. 2791533

22. FREE FLIR: Thermal Dataset for Algorithm Training. https://www.flir.in/oem/adas/adas-dat aset-form

23. Ariffin, S.M., Jamil, N., Rahman, P.N.: DIAST variability illuminated thermal and visi-ble ear images datasets. In: Proceedings of Signal Processing, Algorithms, Architecture, Arrangements, Application (SPA), September 2016, pp. 191–195 (2016)

24. Li, M., Zhang, T., Cui, W.: Research of infrared small pedestrian target detection based on YOLOv3. Infr. Technoiogy 42(2), 176–181 (2020)

25. Li, Y., Li, S., Du, H., Chen, L., Zhang, D., Li, Y.: YOLO-ACN: focusing on small target and occluded object detection. IEEE Access 8, 227288–227303 (2020)

26. Cao, Y., Zhou, T., Zhu, X.,Su, Y.: Every feature counts: an improved one-stage detec-tor in thermal imagery. In: Proceedings of IEEE 5th International Conference Computer Communication, (ICCC), December 2019, pp. 1965–1969 (2019)

27. Song, X., Gao, S., Chen, C.: A multispectral feature fusion net-work for robust pedestrian detection. Alexandria Eng. J. **60**(1), 73–85 (2021). https://www.sciencedirect.com/science/article/pii/S1110016820302507

28. Hou, Q., Zhou, D., Feng, J.: Coordinate attention for efficient mobile network design. In: IEEE/CVF Conference on Computer Vision and Pattern Recognition (CVPR), pp. 13708–13717 (2021).https://doi.org/10.1109/CVPR46437.2021.01350

29. Jacob, B., Kligys,S., Chen, B., et al.: Quantization and Training of Neural Networks for Efficient Integer-Arithmetic-Only Inference (2017)

30. Devaguptapu, C., Akolekar, N., Sharma, M.M., Balasubramanian, V.N.: Borrow from any-where: pseudo multi-modal object detection in thermal imagery. In: Proceedings of IEEE/CVF Conference on Computer Vision and Pattern Recognition Workshops (CVPRW), pp. 1–10 (2019)

31. Cao, Y., Zhou, T., Zhu, X., Su, Y.: Every featurecounts: an improved one-stage detec-tor in thermal imagery. In: Proceedings IEEE 5th International Conference on Computer Communication (ICCC), pp. 1965–1969 (2019)

32. Zhang, S., Wen, L., Bian, X., Lei, Z., Li, S.Z.: Single-shot refinement neural network for object detection. In: Proceedings of IEEE/CVF Conference on Computer Vision Pattern Recognition, pp. 4203–421 (2018)

33. Li, S., Li, Y., Li, Y., Li, M., Xu, X.: YOLO-FIRI: Improved YOLOv5 for infrared image object detection. IEEE Access. 9, 141861–141875 (2021)

Slow Port Scanning Attack Detection Algorithm Based on Dynamic Time Window Mechanism

Ming Ying[✉] [iD]

School of Computer Science and Engineering, Tianjin University of Technology,
Tianjin, China
mergn@stud.tjut.edu.cn

Abstract. Scanning attacks based on network protocols such as TCP, UDP and ICMP are varied, and it is difficult to detect slow scanning attacks due to the large volume of normal network traffic. Aiming at the above problems, this paper proposes a slow port scanning attack detection algorithm based on dynamic time window mechanism. First, the scan rate feature of scanning attack is extracted from the sending rate of responses from the victim using the exponentially weighted moving-average (EWMA). Then, the dynamic strategy changes the length of time window according to the scan rate feature, and connection event is generated within the time window. Taking the connection event as a sample, the scanning attack detection is set as a Sequential Probability Ratio Test. Finally, it is determined whether the host is a scanning attacker by the test result. The experimental results show that the method provides 93.9% detection rate and 0.1% false alarm rate, and the detection delay is reduced by 25.3% compared with the detection algorithm based on the fixed time window mechanism.

Keywords: Slow port scanning attack · Dynamic time window · Sequential Probability Ratio Test · EWMA

1 Introduction

Port scanning attack is an information collecting activity that launched before the hackers invading the target network. It provides valuable information of the target network for other intrusion behaviors, such as ransomware, botnet attacks and DDoS attacks [1,2]. The attacker sends scanning probes to the victim using different network protocols. The specific responses that triggered by these probes indicates the status of network services of the victim [3].

The fast scan attack sends a large number of probes in a short period, and it is easy to be detected by the intrusion detection systems [4]. In order to evade the detection and improve the accuracy of the scanning results, the attacker will expand the sending intervals of the scanning probes to slow down the scans [5,6]. Scanning attacks that follow this strategy are regarded as slow port scanning

© The Author(s), under exclusive license to Springer Nature Singapore Pte Ltd. 2023
S. Yang and S. Islam (Eds.): APWeb-WAIM 2022 Workshops, CCIS 1784, pp. 188–201, 2023.
https://doi.org/10.1007/978-981-99-1354-1_17

attacks [7]. Slow port scanning attack is more difficult to detect since the abnormal flows are hidden in the large volume of normal network traffic.

A packet with abnormal flags or a failed network connection between attacker and victim are considered as local features of scanning attacks. The persistence of scanning attacks or statistical values on long-term scale are implies the global features of scanning attacks. Most of the scan detection algorithms extract these two types of features to detect scanning attacks [7]. The threshold random walk algorithm (TRW) [8] has been widely studied. The algorithm focuses on the failed TCP connection to detect TCP scans. TRW algorithm is based on Sequential Probability Ratio Test [3], which introduces a low detection delay. However, if the attacker connects to the open service and the unknown port on victim alternately, the Sequential Probability Ratio Test loses the ability to make a decision. Using time window mechanism can overcome the shortcoming of Sequential Probability Ratio Test [10], but it is difficult to set an appropriate time window length to effectively balance the detection rate and false alarm rate.

Based on the above analysis, this paper proposes a slow port scanning detection algorithm based on dynamic time window mechanism. The main contributions of this paper are the following three aspects:

1. A slow port scan detection algorithm based on dynamic time window.
2. A scan rate estimation method based on exponentially weighted moving average.
3. A dynamic strategy based on the relationship between the scan rate and the length of time window.

2 Related Work

The current research of port scanning detection mainly focuses on the methods based on algorithms, thresholds and soft computing [7]. The fast port scanning attack is completed in a short time period. Snort intrusion detection system [4] is able to detect fast port scan attack, it counts whether the host accesses more than M different destination hosts or N different destination ports within T seconds to make decision. For slow scanning attacks, methods based on fixed threshold is hard to set the appropriate threshold in advance.

Nisa et al. [11] aiming at filtering the network communication patterns that conform the known scanning activity to detect scanning attack. Similarly, Griffioen et al. [12] searches for probes with the same or correlated embedded information to identify scanning programs. The detection rate of these two methods depends on the ability to identifying the communication patterns or the embedded information of probes. Dabbagh et al. [13] proposed a scan detection method to detect slow port scanning attack. The proposed method split the network data packets in fixed time windows, and counts the number of TCP flags during multiple TCP connections to detect abnormal behaviors. The above methods rarely focus on the response messages from the victim.

Jung et al. [8] proposed the threshold random walk algorithm (TRW) [9]. TRW marks the TCP three-way handshake between the remote hosts and local hosts, and then the Sequential Probability Ratio Test model is applied to determine the identity of the remote host, and only 4 or 5 connection attempts are required to make decision. However, TRW algorithm is limited to the complicated connection attempts of attacker [14]. Ring et al. [10] generates network events within a time window, and detects slow scanning attacks using TRW. The time window mechanism is useful to decrease the detection delay. Sekar et al. [15] improved the Snort using different length of time window and proposed a multi-resolution detection method. However, requirement of computing resources grows rapidly with the number of time windows increases. Fukuda et al. [16] uses the Hough transform to identify the scanning rate of port scans, and detect attackers based on discovered speed feature.

3   Features of Scanning Activity

Scanning attackers are more likely to connect with non-existent hosts or closed services compared with normal host [8]. Therefore, scanning activities can be effectively disclosed from the large volume of network traffic by marking the failed network connections. In addition, due to the strategy of scanning programs, port scanning activities have typical scan rate feature [16].

The RST flag in TCP connection is used to abort the abnormal TCP connections. The number of failed TCP connection can be indicated by recording the number of TCP-RST packets. Furthermore, the number of probes sent by scanning attackers in a time period reflects the scan rate of scanning attacks. These two features are calculated in the CIDDS-001 network intrusion detection dataset [17]. The dataset includes normal network communications and different types of port scanning attacks. Figure 1(a) shows the cumulative sum of the TCP-RST packets received by the two normal hosts and the scanning attacker in the time period from 7:25 to 14:00 on the second day of the second week. The scanning attack using speed option T1 (slow scan) is executed at 8:00 and the scanning attack using speed option T2 (slow scan) is executed at 12:00.

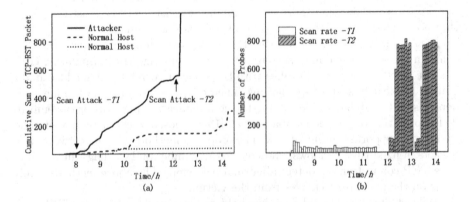

Fig. 1. Features of scanning attack

As shown in figure, the attacker received a large number of TCP-RST packets in both two period. Figure 1(b) shows the number of probes sent by attacker in every 5 min during different scan rate of attacks. In the T1 scan rate mode, an average of 30 probe packets are sent every 5 min (1 probe in 10 s), while the number of probes is maintained at about 800 during the execution of the T2 scan rate mode attack. It can be seen that from the attacker's point of view, the scanning attack activity has a certain scan rate. The scanning attacks in the dataset and the scanning tools used by the attackers are discussed in Sect. 5.

The proposed method selects the following features to detect slow port scanning attacks:

1. The number of failed connections or abnormal requests.
2. Stable scan rate of the scanning attack.
3. Time duration of scanning attack.

4   Proposed Approach

4.1   Marking Suspicious Flows

It is difficult to effectively mark scanning probes, while the victim usually replies with a specific packet to indicate the state information of its service. This work aims to detect three types of scans: TCP SYN scan, UDP scan and ICMP Ping scan [3]. Table 1 lists the response messages that scanning attacker received from the victim when the attacker sends different probes. It is clear that most of the response messages are limited into TCP-RST packet or ICMP error message when the attacker attempts to connect with the non-active host or closed ports. Therefore, it is convenient to locate scanning activities by marking these response messages.

Two rules are proposed to mark suspicious flows for TCP scans, UDP scans and ICMP Ping scans:

1. For TCP scans, if the protocol type of a flow data is TCP and cumulative sum of flags contains the RST flag and does not contain the PSH, FIN and URG flag, then the flow is marked as suspicious flow.
2. For UDP scans and ICMP Ping scans, if the protocol type of a flow data is ICMP and the type field is 3 (error message), then the flow is marked as suspicious flow.

Note that the detection method should focus on the destination address which implies the attacker.

4.2   Dynamic Time Window

Exponentially Weighted Moving Average. Exponentially weighted moving average (EWMA) [18] is a time series analysis method. It provides the estimation of the time series by assigning different weights to the data of the time series

Table 1. Response messages received by attacker

Port scanning technique	Open port on active host	Closed port on active host	Non-active host
TCP SYN scan	TCP-SYN	TCP-RST	-
TCP Connect scan	TCP-SYN	TCP-RST	-
UDP scan	UDP Response	ICMP Error	-
ICMP Ping scan	ICMP Response	ICMP Response	ICMP Error

suspicious flow contains the attribute of timestamp to record the start of the flow. Hence, during a scanning attack, D-value sequence of the suspicious flow's timestamp can be calculated. The probe intervals set by the attacker can then be estimated by predicting the latest data of observed D-value sequence.

Assume that the timestamp sequence of suspicious flows received by the attacker is

$$T_i, (i = 0, 1, 2, ...).\tag{1}$$

D-value sequence of T_i can be calculated as

$$D_j = T_j - T_{j-1}, (j = 1, 2, 3, ...).\tag{2}$$

Set an Exponentially Weighted Moving Average Model of D_j to estimate its latest value

$$\begin{cases} Z_0 = \mu_0 \\ Z_j = (1 - r) * Z_{j-1} + r * D_j, (j = 1, 2, 3, ...) \end{cases}.\tag{3}$$

where μ_0 is the initial of the EWMA value and r is the weighting factor. Higher value of r means that the recent data has more impact on the prediction value. When the new suspicious flow T_i arrives, the prediction of time interval of probes is Z_i.

Changing Time Window Using Dynamic Strategy. The length of the time window varies linearly with the estimation of the probe intervals. Define the sequence of time window length as

$$\begin{cases} W_0 = Z_0 \\ W_i = k * Z_i & W_{min} \leq Z_i \leq W_{max}, i > 0 \\ W_i = k * Z_{max} & Z_i > W_{max}, i > 0 \\ W_i = k * Z_{min} & Z_i < W_{min}, i > 0 \end{cases}.\tag{4}$$

The initial length of the dynamic time window is W_0. After calculating the estimation of probe interval Z_i, the length of the time window that bound to the specific host is updated by the new value W_i. The functionality of dynamic factor k is discussed in Sect. 4.4. When the estimated value of the probe interval is too high or too low, the time window is adjusted according to the upper limit W_{max} and the lower limit W_{min}.

4.3 Scanning Detection Method Based on Sequential Probability Ratio Test

After marking the suspicious flows received by attacker and set the time window sequence using EWMA, all the suspicious flows triggered by the attacker are split with the different length of time window. Then, generate a connection event in every single time window which includes zero or numbers of suspicious flows. A connection event contains multiple attributes, as shown in Table 2. The IP is the destination address of suspicious flow. The attribute Rec-Type_1 and Rec-Type_2 counts the number of two types of suspicious flows that received by the IP in current time window. The attribute α is the sum of Rec-Type_1 and Rec-Type_2 and the connection event is classified as success or failure. If the IP of connection event does not receive any suspicious flow data within a time window, the connection event is considered to be a successful connection event, marked as $Y_i = 0$. Otherwise mark the connection event as failure, i.e. $Y_i = 1$, as shown in Eq. 5.

$$\begin{cases} Y_i = 0, \alpha = 0 \\ Y_i = 1, \alpha > 0 \end{cases} \tag{5}$$

Table 2. Attributes within a connection event

Name	Description
IP	IP address related to connection event
Rec-Type_1	The number of received Type_1 suspicious flows
Rec-Type_2	The number of received Type_2 suspicious flows
α	Sum of Rec-Type_1 and Rec-Type_2

The Sequential Probability Ratio Test is a statistical test method [9]. First, set two hypotheses H_0: the IP of connection event is normal host, H_1: the IP of connection event is scanning attacker. Then, given the conditional probability from Eq. 6 to Eq. 9, likelihood ratio for IP can be updated using Eq. 10 to determine which hypotheses should be selected.

$$P[Y_i = 0|H_0] = \theta_0 \tag{6}$$

$$P[Y_i = 1|H_0] = 1 - \theta_0 \tag{7}$$

$$P[Y_i = 0|H_1] = \theta_1 \tag{8}$$

$$P[Y_i = 1|H_1] = 1 - \theta_1 \tag{9}$$

Equation 6 represents the success probability of connection event for a normal host, and Eq. refathe represents the success probability of connection event for a scanning attacker.

$$ratio_i^{IP} = ratio_{i-1}^{IP} * \alpha_i * \frac{P[Y_i = \delta|H_1]}{P[Y_i = \delta|H_0]} \tag{10}$$

The factor α_i in Eq. 10 is the attribute α of i-th connection event, and it takes different value according to conditions in Eq. 11. The initial value of $ratio_i^{IP}$ is defined as 1 and δ is the observed value of Y_i.

$$\begin{cases} \alpha_i = 1, Y_i = 0 \\ \alpha_i = \alpha, Y_i = 1 \end{cases} \tag{11}$$

In order to avoid permanently judging the scanning attacker as a normal host [14], the lower threshold of the Sequential Probability Ratio Test is removed. Finally, if $ratio_i^{IP} \geq \eta$, accept H_1 and the scanning attack alarm is triggered, where η is the upper threshold of the test. Otherwise, no decisions are made and $ratio_i^{IP}$ continues to update until it exceeds the upper threshold.

4.4 Dynamic Adjustment of Time Window

Assuming that the estimated value of the probe interval of an unknown IP is R_0 in average. When the current length of time window is adjusted to W, then there are $\overline{\alpha}$ suspicious flows in per time window in average, as shown in Eq. 12.

$$\overline{\alpha} = \frac{W}{R_0} \tag{12}$$

From Eq. 4, when the dynamic factor $k < 1$, it means that time window is smaller than the estimated value of the probe interval. According to Eq. 5 to 11, when there are enough suspicious flows, at least one successful connection event will be generated, which will reduce the likelihood ratio. Therefore, there is a possibility that the unknown IP will not be marked as a scanning attacker. If the real identity of the unknown IP is the scanning attacker, setting $k < 1$ will reduce the detection rate. Meanwhile, if the real identity of the IP is a normal host, setting $k < 1$ will reduce the false positive rate.

When $k > 1$, it means that the dynamic time window is larger than the estimated value of the probe interval. In this case, consecutive failed connection events will be generated. If the real identity of the unknown IP is the attacker, then setting $k > 1$ will bring higher detection rate, but at the same time it causes higher false positive rate.

Therefore, dynamic factor k can balance the detection rate and false positive rate. Experimental results of different k value is discussed in Sect. 5.5.

5 Experimental Results and Analysis

5.1 Experimental Dataset and Evaluation Criteria

CIDDS-001 [17] is a dataset designed for evaluating intrusion detection systems. The dataset simulates a small enterprise network in an OpenStack environment, generating flow-based network data, which includes various network attacks as well as normal network traffic. Network topology is shown in Fig. 2, it includes external Internet and internal simulation network.

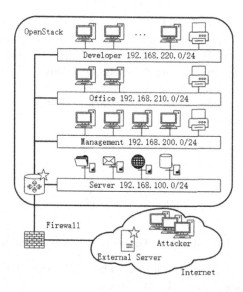

Fig. 2. Topology diagram of experimental environment

NetFlow [19] agents (highlighted by starts) are set in the external and the internal network to collect network traffic. Since most of the external network data is unlabeled, the labeled flow data collected by internal agents are selected. 40 scanning attacks are launched by 2 different hosts in the developer subnet of the internal network using Nmap scanning tool in first and second week [20]. Scanning attackers are allowed to select scanning parameters such as scan rate or probe types. The details of attack traffic have shown in Table 3.

Table 3. Details of scan attack traffic

Port scanning technique	Proportion of scanning traffic /%		
	Scan rate -T1	Scan rate -T2	Scan rate -T3
TCP SYN scan	0.196	0.317	0.296
UDP scan	0.008	0.001	<0.001
ICMP Ping scan	0.016	0.020	0.011

The T3 (probe interval is adjusted by the performance of the machine) option in the Nmap is a fast scan attack, and the T1 (probe interval is 15 s) and T2 (probe interval is 0.4 s) options set the scanning activity as a slow port scan.

The flow data sent by victim is defined as positive, since the proposed method focuses on the response message of the scanning attack rather than the scanning probes. Other flows (including the flows related to the probe sent by the attacker) are defined as negative. Three evaluation criteria are used to evaluate the performance of proposed scan detection algorithm.

1. The true positive rate (TPR), also named as detection rate, is the ratio of the correctly detected positive samples to all the positive samples.
2. False alarm rate (FAR), which is the ratio of the detected positive samples (negative in real) to all the negative samples. The TPR and FAR are calculated according to Eq. 13.
3. Detection delay, which is the time duration between the start of the scan attack and the moment of detection.

$$
\begin{cases}
TPR = \dfrac{TP}{TP+FN} \\
FPR = \dfrac{FP}{FP+TN}
\end{cases}
\tag{13}
$$

5.2   Experimental Results of Suspicious Flow Marking Rules

Table 4 shows the details of scanning attacks during the time period from 13:30 to 16:30 on the second day of the first week.

Table 4. Time duration of scanning attack

Port scanning technique	Scan rate	Start time	End time
TCP SYN scan	T2	13:52	14:56
ICMP Ping scan	T2	15:05	15:20
TCP SYN scan	T1	15:27	20:23

11274 flows (about 3.58%) marked as suspicious and 313918 flows are regarded as non-suspicious flows. As shown in Table 5, the marking rule is evaluated by the TPR and FAR, since not all of the suspicious flows are related to scanning attack and not all the non-suspicious flows are normal network traffic.

Table 5. Experimental result

TPR /%	FAR /%
95.81	1.11

The results show that the marking rules can filter a large volume of normal flows. When the victim opens the port that scanned by the attacker, a successful connection between attacker and victim will be established. Therefore, it is difficult to identify 4.18% of the response flows from the victim as suspicious flows, which reduces TPR of the proposed approach.

5.3   Experimental Results of Scan Rate Estimation

During the time period as selected in Sect. 5.2 from the dataset, both normal host and scanning attacker received suspicious flow data and the scan rate is estimated. Since the normal host do not send probes and usually access the known service on active host, they rarely receive suspicious flows. As shown in Fig. 3, scan rate is estimated on both normal host and attacker.

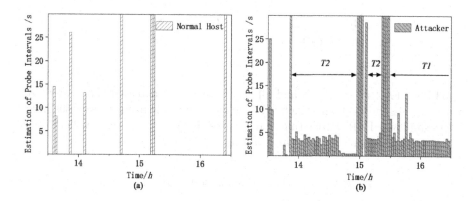

Fig. 3. Estimation of scan rate

The Fig. 3 shows that the estimated value of the probe interval is updated when new suspicious flow data is identified. Normal host has low frequency of estimating the probe intervals while attacker's probe intervals are estimated in high frequency, because normal host rarely receive suspicious flows. Figure 3(b) shows that during the T1 scan, the probe intervals are maintained within 15s and the value is maintained within 6s in T2 scan.

5.4   Experimental Results of Port Scan Attack Detection Method

The proposed approach is compared with three other methods, namely Snort [4], the slow scanning detection method developed by Nisa et al. [11] and detection method based on fixed time window mechanism. Detection method based on fixed time window means the length of time window is taken as a fixed value, as shown in Eq. (14).

$$W_i = W_f, (i = 1, 2, 3, ...) \tag{14}$$

Four detection methods are configured through the parameter values in Table 6. Four weeks of flow data from CIDDS-001 is set as input data and the results are shown in Table 7. As shown in table, Snort can detect 34 scanning attacks, but higher false alarm rate and detection delay means that it cannot detect slow scans effectively. The previous approach [11] is designed to detect TCP scans. Therefore, its detection rate is limited and it is able to detect only 25 scanning activities. However, the TPR of proposed approach in this paper

Table 6. Experimental parameter setting

Detection method	Parameter name	Parameter value
Dynamic Time Window	μ_0, r	60(s), 0.2
	θ_0, θ_1, η	0.8, 0.2, 99
	W_{min}, W_{max}, k	3(s), 1800(s), 1
Fixed time window	θ_0, θ_1, η	0.8, 0.2, 99
	W_f	45(s)
Snort [4]	T, N, M	300(s), 22, 22
Previous approach [11]	W	20(s)

is 93.99%, FAR is only 0.1% and all the 40 scans are detected. Compared with the previous approach and the fixed time window mechanism, it provides higher TPR and lower FAR. Furthermore, it reduces the detection delay by 25.3% compared with fixed time window mechanism.

Table 7. Experimental results of detection methods

Detection method	TPR/%	FAR/%	Detection delay/s	Detected activities
Dynamic time window	93.99	0.10	335.2	40
Fixed time window	92.99	0.11	449.4	40
Snort [4]	85.54	45.75	2132.5	34
Previous approach [11]	90.44	0.12	704.7	25

5.5 Experimental Results of Dynamic Factor's Functionality

The dynamic factor k can balance the detection rate and false alarm rate as discussed above. The different curves in Fig. 4 depict the changes of TPR and FAR when dynamic factor k takes different values. Higher value of k can increase TPR and FAR, but the TPR limited near 94% when FAR continues to increase. The complicated scanning activities, such as changing the scan rate or changing the probe type are responsible for the certain results. Figure 5 depicts the change of average detection delay and the number of detected attacks. The smaller time window provides lower detection delay, but it is difficult to detect the 40 attacks. On the contrary, setting higher value of k causes higher detection delay. It can be found that the overall performance of the proposed approach is optimal when $k = 1$.

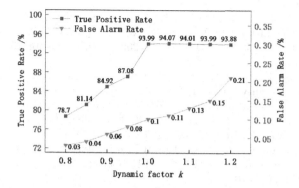

Fig. 4. Impact of dynamic strategy on TPR and FAR

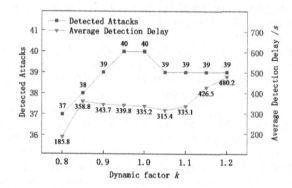

Fig. 5. Impact of dynamic strategy on detected attacks and detection delay

6   Conclusion

A slow port scan detection algorithm based on dynamic time window is proposed. Port scanning attacks usually causes a large number of failed connections. Based on this, suspicious flows are marked, and scan rate is effectively estimated using EWMA from the sequence of suspicious flows. Then, the connection events are generated, and Sequential Probability Ratio Test is applied to detect scanning attacks. The scan detection algorithm based on dynamic time window is able to balance the TPR and FAR. Furthermore, the proposed approach provides a lower detection delay compared with the detection method based on fixed time window mechanism.

References

1. Caroscio, E., Paul, J., Murray, J., Bhunia, S.: Analyzing the ransomware attack on D.C. metropolitan police department by babuk. In: 2022 IEEE International Systems Conference (SysCon), pp. 1–8. IEEE, Canada (2022) https://doi.org/10. 1109/SysCon53536.2022.9773935

2. Hussain, F., Abbas S.G., Pires, I.M., Tanveer, S., Fayyaz, U.U., Garcia, N.M., et al.: A two-fold machine learning approach to prevent and detect IoT botnet attacks. IEEE Access **9**, 163412–163430 (2021) https://doi.org/10.1109/ACCESS.2021.3131014

3. Bou-Harb, E., Debbabi, M., Assi, C.: Cyber scanning: a comprehensive survey. IEEE Commun. Surv. Tutor. **16**(3), 1496–1519 (2013)

4. Roesch, M.: Snort - lightweight intrusion detection for networks. In: Proceedings of the 13th Conference on Systems Administration (LISA-1999), pp. 229–238. (1999)

5. Zhang, Z., Towey, D., Ying, Z., Zhang, Y., Zhou, Z.: MT4NS: metamorphic testing for network scanning. In: 2021 IEEE/ACM 6th International Workshop on Metamorphic Testing (MET), pp. 17–23. IEEE, Spain (2021) https://doi.org/10.1109/MET52542.2021.00010

6. Singh, R.R., Tomar, D.S.: Network forensics: detection and analysis of stealth port scanning attack. Int. J. Comput. Netw. Commun. Secur. **3**(2), 33–42 (2015)

7. Bhuyan, M.H., Bhattacharyya, D.K., Kalita, J..K.: Surveying port scans and their detection methodologies. Comput. J. **54**(10), 1565–1581 (2011) https://doi.org/10.1093/comjnl/bxr035

8. Jung, J., Paxson, V., Berger, A.W., Balakrishnan, H.: Fast portscan detection using sequential hypothesis testing. In: IEEE Symposium on Security and Privacy, pp. 211–225 (2004) https://doi.org/10.1109/SECPRI.2004.1301325

9. Wald, A.: Sequential Analysis. John Wiley & Sons, New York (1947)

10. Ring, M., Landes, D., Hotho, A.: Detection of slow port scans in flow-based network traffic. PLoS ONE **13**(9) (2018) https://doi.org/10.1371/journal.pone.0204507

11. Nisa, M.U., Kifayat, K.: Detection of slow port scanning attacks. In: 2020 International Conference on Cyber Warfare and Security (ICCWS), pp. 1–7. IEEE, Pakistan (2020) https://doi.org/10.1109/ICCWS48432.2020.9292389

12. Griffioen, H., Doerr, C.: Discovering collaboration: unveiling slow, distributed scanners based on common header field patterns. In: NOMS 2020–2020 IEEE/IFIP Network Operations and Management Symposium, pp. 1–9. IEEE, Hungary (2020) https://doi.org/10.1109/NOMS47738.2020.9110444

13. Dabbagh, M., Ghandour, A.J., Fawaz, K., Hajj, W.E., Hajj, H.: Slow port scanning detection. In: 2011 7th International Conference on Information Assurance and Security (IAS), pp. 228–233. IEEE, Malaysia (2011). https://doi.org/10.1109/ISIAS.2011.6122824

14. Mell, P., Harang, R.: Limitations to threshold random walk scan detection and mitigating enhancements. In: 2013 IEEE Conference on Communications and Network Security (CNS), pp. 332–340. IEEE, USA (2013)

15. Sekar, V., Xie, Y., Reiter, M.K., Zhang, H.: A multi-resolution approach for worm detection and containment. In: International Conference on Dependable Systems and Networks (DSN'06), pp. 189–198. IEEE, USA (2006). https://doi.org/10.1109/DSN.2006.6

16. Fukuda, K., Fontugne, R.: Estimating speed of scanning activities with a hough transform. In: 2010 IEEE International Conference on Communications, pp. 1–5. IEEE, South Africa (2010)

17. Ring, M., Wunderlich, S., Grüdl, D., Landes, D., Hotho A.: Flow-based benchmark data sets for intrusion detection. In: European Conference on Cyber Warfare and Security (ECCWS), pp. 361–369 (2017)

18. Roberts, S.W.: Control chart tests based on geometric moving averages. Am. Soc. Qual. Control Am. Stat. Assoc. **42**(1), 97–101 (2000)
19. Cisco systems netflow services export version 9. https://icm.krasn.ru/ftp/rfc/rfc3954.pdf. Accessed 6 June 2022
20. Lyon, G.F.: Nmap network scanning: the official NMAP project guide to network discovery and security scanning. Nmap Project, US (2009)

AU-GAN: Attention U-Net Based on a Built-In Attention for Multi-domain Image-to-Image Translation

Caie Xu[1(✉)], Jin Gan[2], Mingyang Wu[1], and Dandan Ni[3]

[1] Zhejiang University of Science and Technology, Hangzhou, China
caiexu@163.com
[2] University of California, San Diego, USA
j6gan@ucsd.edu
[3] Zhejiang University, Hangzhou, China
nidd@zju.edu.cn

Abstract. Multi-domain image-to-image translation refers to mapping images from a source domain to multiple target domains. The state-of-the-art deep learning models are able to learn mappings among multiple domains without paired image data. The key to most multi-domain image-to-image translation tasks, for instance, object transfiguration, facial attribute editing, etc., is to detect the different regions of interest in the source domain and to convert them from the source domain to different target domains. However, to our knowledge, existing approaches have limited scalability and robustness in detecting the most discriminative semantic part of images in the task of multi-domain image-to-image translation. To address this limitation, we propose Attention U-Net Generative Adversarial Network (AU-GAN), which utilizes three well-known networks (the Attention Gate Network, the Recurrent Residual Convolutional Neural Network (RRCNN) as well as the built-in attention mechanism) for the first time. Experiments indicated that the proposed model not only detects different regions of interest but also achieves figure-ground segregation in the task of image-to-image translation. The proposed models are tested on two benchmark datasets (CelebA, CelebA-HQ). Both qualitative and quantitative results demonstrated that our approach is effective to generate sharper and more accurate images than existing models.

Keywords: Image-to-Image Translation · Attention Gate Network · Residual Convolutional Neural Network · AU-GAN

1 Introduction

Recently, deep learning based on Deep Convolutional Neural Networks (DCNN) has shown great success in the field of computer vision. For example, without

Supported by organization x.

designing handcrafted features, image classification models based on DCNN [1–3] have achieved better performance than handcraft features by a significant margin on many benchmark datasets [4,5]. Deep learning based approach (DCNN in particular) provides state-of-the-art performance for various computer vision tasks for several reasons: first, activation functions and several efficient optimization techniques have been proposed to resolve training problems in DL approaches to a great extent; Second, a deep learning technology named residual learning greatly eases the training of deeper convolutional neural networks [6]; Third, DCNN naturally integrates low, mid and high-level features and the "levels" of features can be enriched with the increase of the number of stacked layers [2]. Because of these three factors, most researchers trying to implement bigger and deeper DCNN architectures like Transformer [7] or a Residual Network with 1001 layers [8] to achieve higher accuracy on different benchmark datasets. However, with the development of larger and deeper deep learning models, while the DCNN provides state-of-the-art accuracy in various computer vision tasks, new failures such as the training of DCNN becoming more complex and expensive are born.

Instead of improving performance in various tasks with bigger and deeper DCNN models, the concept of Recurrent Convolution Layers (RCLs) [9] is introduced in a lot of DCNN models recently. Among them the most representative one is the Recurrent Convolutional Neural Network (RCNN) which contains several blocks of RCLs followed by a max-pooling layer and achieved state-of-the-art accuracy for object classification at that time [10]. Subsequently, RCNN combined with other deep learning models has received widespread attention. For instance, the Long-term Recurrent Convolutional Network (LRCN) was proposed for visual recognition. It consists of two networks: Convolutional Neural Network (CNN) is used for feature extraction; Long Short-Term Memory (LSTM) is applied to observe how features vary with respect to time [11]. [12] proposed a new DCNN model called the Inception Recurrent Residual Convolutional Neural Network (IRRCNN), which utilizes the Recurrent Convolutional Neural Network (RCNN), the Inception network, and the Residual network for image classification. The R2U-Net [6] integrates the RCNN, Residual Network and U-Net for image segmentation and shows superior performance on segmentation tasks compared to equivalent models including U-Net and residual U-Net.

RCNN is not applied in any image translation models to the best of my knowledge. I conjecture that a possible cause is that a lot of attention is paid to how to make Generative Adversarial Networks (GANs) [13] work well in image translation since it was introduced into image translation and has achieved outstanding performance in image-to-image translation. For example, by introducing adversarial learning, Pix2pix GAN [14] greatly improves the quality of image translation. By making the DCNN model learn the mapping from one image domain to another without paired data, Cycle GAN [15] proposed a novel loss function called cycle consistency loss. Most previous GAN models based on DCNN were easy to produce unwanted changes such as the background of generated images being easily affected. In order to address these limitations, the Contrast GAN [16] was proposed, which first crops a part in the image according to the masks, then makes translations, and finally pastes it back. Although it has obtained promising results, it is hard to collect training data with object masks. Instead of using the object masks directly, Attention GAN [17] is used to train an extra model to produce the object masks and fit them into the generated image patches. Because the number of parameters of a special DCNN model used to produce the object masks is tremendous, the training complexity in both time and space would increase. Subsequently, the Attention-Guide GAN [18] (AG-GAN) was proposed, which can produce object masks via a built-in attention mechanism without using extra data and DCNN models. However, the object masks generated from AG-GAN are dissatisfactory.

Rather than compressing an entire image or a sequence into a static representation, attention mechanisms allow a model to focus on the most relevant part of images [17]. Due to their advantage and theoretical support, attention mechanisms have been gaining considerable attention, and the desire to use attention mechanisms in many fields is growing. [19] proposed a self-attention mechanism in GANs, which it is conducive to drawing image features in which fine details at every location are carefully coordinated with fine details in distant portions of the image. [20] proposed a Dual Attention Network (DA-Net) for Scene Segmentation, which can adaptively integrate local features with global dependencies. However, the attention mechanisms mentioned above are used only for a certain feature map. From an architectural point of view, the DCNN model for image translation tasks requires both convolutional encoding and decoding units. Subsequently, a novel Attention Gate [21] was proposed to act on encoding and decoding units at the same time, which can implicitly learn to suppress irrelevant regions in an input image while highlighting salient features useful for a specific task.

In this paper, we are presenting an improved version of the DCNN model of image translation inspired by the recently developed promising DCNN technologies like the RRCNN, the Attention Gate, the U-Net [22] and the Built-in Attention mechanism. The proposed model not only ensures a better quality of image translation with the same number of network parameters against other DCNN architectures but also helps to improve the convergence speed. The contributions of this work are as follows:

- A novel framework of generator based on DCNN was proposed, which fuses RRCNN, U-Net, and Attention Gate.
- The experiment indicated that our model is superior to the baseline models including AG-GAN and Star-GAN on two famous datasets.
- The experiment showed that the proposed model is conducive to quickening learning speed and ensures better feature representation for image translation tasks.

2   Related Work

2.1   Generative Adversarial Networks (GANs)

GAN is a type of neural network architecture for the generative model which consists of two sub-networks: Generator G which captures the data distribution and the Discriminator D which estimates the probability that a sample comes from the training data rather than Generator G. Generator G and Discriminator D are trained in adversarial ways. The goal of Generator G is to maximize the probability of D making a mistake, while Discriminator D is to differentiate the generative samples from real samples as much as possible. Two sub-networks contest with each other in the sense of game theory. This process can be described as:

$$\frac{min}{G}\frac{max}{D}E_{x\,p_{data}}[logD(x)] + E_{z\,p_z}[log(1 - D(G(z)))] \tag{1}$$

GAN was originally used to generate images. However, due to its theoretical and practical success, GAN has been introduced to various other tasks. As far as image translation is concerned, Pix2pix is a conditional framework using a CGAN to learn a mapping function from input to output images with paired data. Cycle GAN learns the mappings between two image domains without the paired images. Star GAN [23] provides a remarkable benchmark architecture for multi-domain image-to-image translation, which is able to learn a mapping function from one image domain to others with unpaired image data.

Attention-Guided Image-to-Image Translation. The task of image-to-image translation aims to translate images from a source domain to another one or more target domains. Instead of taking the image as a whole to accomplish the transformation, most image translation tasks aim to transform a particular type of object in an image into another type of object without influencing the background regions. To explain this phenomenon using the task of facial attribute editing as an example, facial attribute editing aims to manipulate single or multiple attributes of a face image, i.e., to generate a new face with desired attributes while preserving other details. However, we can find that the girl's hair color in the image is transformed into other colors and the background color is changed as well in the top line of Fig. 1. In order to fix the aforementioned limitations, Contrast GAN was proposed, which disentangles the image background with object mask changes by exploiting the semantic annotations in both the train and test phases. However, the collection of object masks is often expensive and time-consuming. Therefore, Attention GAN was proposed, which uses two parallel generative networks, in which one is used to generating an object mask. However, since the introduction of the extra attention network to generate object masks, the number of parameters has become huge. All these aforementioned methods employ extra networks to generate object masks, which would increase both the number of parameters and the training time of image translation models. [18] proposed a novel Network (AGGAN), which can detect the most discriminative semantic object and minimize changes of unwanted parts for semantic manipulation problems without using extra data and models. However, the quality of object masks generated from AGGAN is often poor. Recently, the such architecture of generative models that utilizes object masks has been widely concerned in the field of image translation.

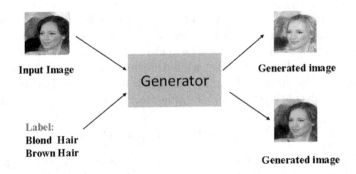

Fig. 1. An example to explain that the background regions will be influenced in face attribute editing.

2.2 U-Net and Its Variants

Initially, "U-Net" is one of the very first and most popular approaches for semantic medical image segmentation. The network of the basic U-Net consists of two parts: the convolutional encoding and decoding units, where the convolutional encoding unit is used to extract semantic features of images, and the convolutional decoding unit is used to produce segmentation maps. Because of U-Net's outstanding performance, different variants of it have been proposed. [21] is among the most famous U-Net variants, which proposed a novel attention gate (AG) model. Models trained with AGs would implicitly learn to suppress irrelevant regions in an input image while highlighting salient features useful for a specific task. [6] proposed a Recurrent Residual Convolutional Neural Network (RRCNN) based on U-Net models for image segmentation, which utilizes the power of U-Net, Residual Network, as well as RCNN. Although U-Net and its variants have made extraordinary progress in the field of image segmentation, it has been rarely used in the field of image translation. In this paper, we designed a novel generative model for image translation, which combined the power of the Attention Gate and RRCNN.

3   Method

In this section, we will introduce the overview of AU-GAN and its training objective functions. Next, we first start with the Attention U-Net generator with RRCNN and discriminator of the proposed Attention U-Net Generative Adversarial Network (AU-GAN). And then we will introduce the loss function for better optimization of the model. Finally, we will present the implementation details of the whole model including network architecture and training procedures.

3.1   An Overview of AU-GAN

First, let $x_i \in X$ and $y_i \in Y$ denote the training images in source and target image domain respectively. Our model aims to train a single generator G that not just learns mappings among multiple domains without paired data but also detects the object of interest and produces the corresponding object mask. In order to achieve these two functions, a novel generator with the Attention U-Net, RRCNN and a built-in attention mechanism was introduced in the paper.

More specifically, instead of generating the image from the generator directly, the generator of Attention U-Net outputs four channels, which can be divided into two parts: the first channel produces the object mask, which represents the background of target domain y_B; the next three channels output the foreground of target domain y_m. The final generated image was obtained by the following formulation:

$$y = y_B * y_m + (1 - y_B) * x \tag{2}$$

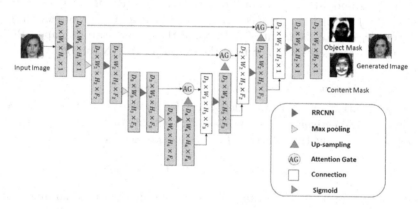

Fig. 2. Overview of the Generator of AU-GAN, consisting of four parts: RRCNN, Max pooling, Up-sampling unit and Attention Gate.

3.2   Attention U-Net Generator with RCL for Image Translation

The differences between the proposed models with respect to the previous models are three-fold. Firstly, instead of using regular forward convolutional layers in both the encoding and decoding units, the RRCNN was used. There are two benefits to this image translation task: one is that the RRCNN improves the convergence of generator loss to facilitate the training of the generator, and the other is that the efficient feature accumulation method is included in the RCL units of both proposed models. Secondly, the Attention Gate was incorporated into the standard U-Net architecture to highlight salient features that are passed through the skip connections instead of using the skip connections directly. Lastly, a novel generator architecture with a built-in attention mechanism can not only detect the most discriminative semantic part of images but also produce the object mask in different domains at the same time. Next, I'll introduce RRCNN and Attention Gate Network in detail.

3.3   Recurrent Residual Convolutional Neural Network (RRCNN)

RRCNN utilizes the strengths of two recently-developed deep learning models, which are the deep residual network that is famous for improving the convergence of deeper neural networks and the Recurrent Convolutional Neural Network (RCNN) that can enhance the ability of the model to integrate the context information respectively. The architecture of RRCNN is shown on the left of Fig. 3, which includes two main parts: the standard skip connection layers and the Recurrent Convolutional Neural Network. The key module of RRCNN is the RCNN unit. The unfold operations of the RCNN are shown on the right of Fig. 3, which includes two main parts: the standard convolution layer and the Recurrent Convolutional Layers (RCL). The feature map of RCNN is first fed into the standard convolution layer, and then the Recurrent Convolutional Layers (RCL) perform with respect to the discrete time steps. More formally, let x_{input} and x_t denote the input of RCNN and the output feature map of RCNN at time step t respectively. Therefore, the output of RCNN can be expressed as follows:

$$x_{output} = x_{input} * w_{input} + (x_{t-1} + x_{input}) * w_r + B_r \tag{3}$$

In the equation, w_{input} and w_r denote the feedforward weights and the recurrent weights respectively. B_r is the bias. The outputs of RCNN are fed to the standard ReLU activation function f and are expressed as:

$$x_{output} = f(x_{output}) = max(0, x_{output}) \tag{4}$$

The output of both RCNN units is passed through the residual unit that is shown on the left of Fig. 3. In this paper, the final output of RRCNN is used for down-sampling and up-sampling layers in the convolutional encoding and decoding units of the RU-Net model respectively.

3.4   Attention Gates

The architecture of the Attention Gate is shown in Fig. 4. In this paper, instead of establishing the skip connections between encoding and decoding units like U-Net, Attention Gate was incorporated into the standard U-Net architecture, as shown in Fig. 2. Compared with the standard U-Net architecture, such architecture has the following advantages: First, it can disambiguate irrelevant and noisy responses in skip connections. Second, AGs filter the neuron activations during the forward pass as well as during the backward pass. Gradients originating from background regions are down-weighted during the backward pass. This allows model parameters in shallower layers to be updated mostly based

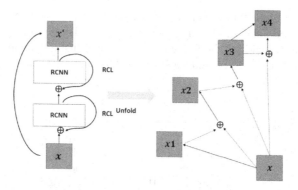

Fig. 3. The left is the architecture of RRCNN and the right is the unfolded recurrent convolutional neural network.

on spatial regions that are relevant to a given task. The Attention Gate has two inputs, one of which is extracted from the encoding unit and the other comes from the decoding unit. Each convolution includes two parts: one is convolution operation and the kernel size and stride size are all 1 * 1, while the other is Batch Normalization (BN).

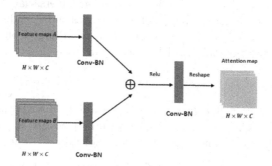

Fig. 4. The architecture of Attention Gate. Feature maps A is obtained from encoding units, while feature map B is obtained from decoding units.

3.5   The CNN-Based Patch Discriminator

As shown in Fig. 5, the CNN-based discriminator is a fully convolutional network (FCN), which consists of eight convolutional neural networks. The first six convolution layers are used for extracting high-level features from the input image accurately and the other two connected layers are used to give a probability that the input image comes from the target domain and the category of the input image respectively. For the whole discriminator network, except the last two convolution layers, the kernel size and stride size of convolutional neural networks are 4 * 4 and 2 * 2 respectively. And the number of output channels has multiplied from 64 to 2048 after each convolutional neural network. The kernel size and the stride size of the last two convolution layers are 3 * 3 and 1 * 1. Each convolution includes the following two operations: a convolution operation and a Leaky ReLU with a negative slope of 0.01.

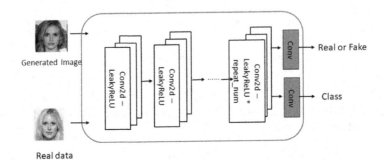

Fig. 5. The architecture of CNN-based Patch Discriminator.

3.6   Optimization Objective

Adversarial Loss: The purpose of adopting an adversarial loss learned from WGAN-GP [24] is to make the generated images indistinguishable from real images. The generator G generates an image conditioned on both the input image and the target domain, which is shown in Eq. 5.

$$L_{adv} = E_x[D_{src}(x)] - E_{x,c}[D_{src}(G(x,c))] \tag{5}$$

In this paper, we define $D_{src}(x)$ as a probability distribution of the source data given by D. The generator G tries to minimize this objective, while the discriminator D aims to maximize it during the training stage.

Classification Loss: Aiming to make the category of generated images correspond to the target label c, the discriminator D has to identify the class of generated and real images in the target domain. Hence, we add an auxiliary classifier on top of the discriminator D. The auxiliary classifier distinguishes the distribution of attributes from labels and others by optimizing the parameter in the discriminator D. Meanwhile, G is updated via backpropagation. The process is defined as Eq. 6.

$$L_{cls} = -E_x log([D_{cls}(c^{\sim} \mid y)]) - E_{x,c}[log(D_{cls}(c \mid G(x,c)))] \tag{6}$$

where the term $D_{cls}(c^{\sim} \mid y)$ represents the probability that the input image x comes from $c^{\sim}$ domain, which is computed by D.

Cycle Consistency Loss: In order to guarantee that the generated image can change only the domain-related part of the inputs while preserving the other content, we introduce a cycle consistency loss into the generator, namely, by taking the generated image X_{fake} and the opposite domain $c \sim$ into the generator, the reconstructed image should be consistent with the corresponding input image X, which is formulated as:

$$L_{rec} = E_{x\ p_{data}, x_{fake}\ p_g}[\|x - G(G(x,c), c^{\sim})\|] \tag{7}$$

Attention Loss: Similar to the cycle consistency loss that requires a sample from one domain to the other that can be mapped back to produce the original sample, the attention loss means that the object mask generated from the built-in attention mechanism should be consistent, whether from source domain to target domain or from target domain to source domain. The attention loss function is defined as:

$$L_{atten} = \|A_i(x) - A_i(G(x))\|_1 \tag{8}$$

Overall Objective: Finally, the objectives for the generator and the discriminator are formulated as Eq. 10 and Eq. 11 respectively.

$$L_D = \mu_1 * L_{adv} + \mu_2 * L_{cls} \tag{9}$$

$$L_G = \mu_1 * L_{adv} + \mu_2 * L_{cls} + \mu_3 * L_{rec} + \mu_4 * L_{atten} \tag{10}$$

μ_1 and μ_2 represent the coefficients of adversarial loss and classification loss of generator and discriminator. μ_3 and μ_4 are coefficients of reconstruction loss and attention loss of generator. In this work, the coefficients of weights are set to 10, 2, 10, and 1 respectively.

4 Experiments

In this section, we first introduce the two datasets used in this paper, the baseline models for comparison and the corresponding metrics to evaluate the quality of generated images. Then, we compare AU-GAN against state-of-the-art methods. Lastly, ablation experiment is used to demonstrate the effect of RRCNN, Attention Gate as well as a built-in attention on image-to-image translation.

4.1 Experimental Setup

Dataset: Two public datasets are used to evaluate the proposed method, which are Large-scale Celeb Faces Attributes (CelebA) dataset [25] and a higher-quality version of the CelebA (CelebA-HQ) dataset [26]. CelebA contains 10,177 identities, 202,599 face images, and 40 binary attribute annotations per image. CelebA-HQ is a higher-quality version of the CELEBA dataset, which has a total of 30K images, 40 binary attribute annotations per image and each with a resolution of

1024 * 1024. In this work, we randomly selected 2,000 and 1,000 images from CelebA and CelebA-HQ respectively for testing and quantitative evaluating, while the rest were used for training.

Baseline Models: We consider two state-of-the-art cross-domain image generation models as our baselines, which are the Star-GAN and the AG-GAN that has achieved great success in Multi-Domain Image-to-Image Translation.

Evaluation Metrics: We choose two well-known metrics for quantitative evaluation: Kernel Inception Distance (KID) [27] and Fréchet Inception Distance (FID) [28]. Though alternatives like the Inception score (IS) [29] and Amazon Mechanical Turk (AMT) exist, they are either inappropriate or costly. The Inception score computes the KL divergence between the conditional class distribution and the marginal class distribution. However, it has serious limitations-it is intended primarily to ensure that the model generates samples that can be confidently recognized as belonging to a specific class and that the model generates samples from many classes, not necessarily to assess realism of details or intra-class diversity. FID and KID are more principled and comprehensive metrics and have been shown to be more consistent with human evaluation in assessing the realism and variation of the generated samples. The former calculates the Wasserstein-2 distance between the generated images and the real images in the feature space of an Inception-v3 network. The latter measures the dissimilarity between two probability distributions, namely the real and the generated samples, using samples drawn independently from each distribution. Lower values of them mean closer distances between synthetic and real data distributions. In all of our experiments, 2k samples are randomly generated for each model, and the same number of samples for the specified attributes are generated for each model.

4.2   Comparison with the State-of-the-Art

In this experiment, we will edit five attributes of the input image, which are 'Black hair', 'Blond Hair', 'Brown hair', 'Gender', and 'Age'. When editing the hair color of the input image, no matter what is the input hair color, the target domains are 'Black hair', 'Blond Hair', and 'Brown hair'. For the other two properties ('Gender' and 'Age' respectively), the target domains would be contrary to the original properties of the input image. For instance, if the input image is male, the target domain would be female and vice versa. The results of the experiment are shown in Fig. 6, and the scores of FID and KID are shown in Table 1. From them, we can find the following phenomena: (1) The proposed model AU-GAN is superior to the comparison models in both qualitative and quantitative evaluations, which proved that our model is effective. (2) The result of the same model is different on different datasets. For example, the face of generated images is not affected by the result in CelebA, but it is influenced in CelebA-HQ when editing the input of hair color. It shows that the number of datasets has a great influence on image-to-image deep learning. (3) Compared with Star-GAN, the background of the generated image was hardly affected by editing hair color.

We hypothesized that the generator of Star-GAN does not have the ability to detect the most discriminative semantic part of images, which thus makes the background of generated images have been affected. (4) It is obvious that the generated object mask and content mask of AU-GAN were far superior to AG-GAN by showing more precise attention. A possible reason for this phenomenon is that the generator of AG-GAN fails to capture semantic features well.

Fig. 6. Generated images from different models on CelebA and CelebA-HQ datasets. The first three lines are the results on CelebA, while the following lines are CelebA-HQ. From top to bottom are Star-GAN, AG-GAN and AU-GAN. From left to right are the input image, "Black hair", "Blond hair", "Brown hair", "Gender" and "Age" in each group.

Table 1. Comparison of the scores of FID and KID of the proposed method and compare models based on database CelebA and CelebA-HQ.

Dataset	CelebA				CelebA-HQ			
	Blond hair		Gender		Blond hair		Gender	
	FID	KID	FID	KID	FID	KID	FID	KID
Star-GAN	25.783	1.246	20.227	0.703	34.152	2.332	31.983	1.942
AG-GAN	17.064	0.930	7.961	**0.172**	20.396	1.119	13.395	0.620
AU-GAN	**16.033**	**0.811**	**7.600**	0.197	**18.802**	**1.040**	**11.905**	**0.391**

4.3 Ablation Experiment

In this paper, we use three novel neural networks, which are RRCNN, Attention Gate, and a built-in attention mechanism respectively. The ablation experiment was performed to verify the impact of each neural network on the image-to-image translation. An important point about RRCNN needs to be highlighted.

As shown in the third section of Sect. 2, the key unit of RRCNN is RCNN. When the step of RCNN (t) is set to 1, the RRCNN would become a plain residual convolution neural network (Fig. 7).

Fig. 7. The generated object mask and content mask of AG-GAN and AU-GAN. Every two lines are a group. From top to bottom, the first is the result of AGGAN and the second is AUGAN.

4.4 The Built-In Attention Mechanism Was Conducive to Separating Face from the Background Image

In order to verify the contribution of the built-in attention mechanism to image-to-image translation, we performed a comparison between AU-GAN without the built-in attention mechanism and the proposed AU-GAN, which is shown in Fig. 8. Apparently, the background of generated images from the proposed AU-GAN is hardly affected. However, the background of synthetic images from AU-GAN without the built-in attention mechanism has obvious changes. Results from what has been discussed above suggest that a built-in attention mechanism can not only improve the quality of generated images but also be conducive to separating the target object from the background image.

4.5 The Recurrent Residual Convolutional Neural Network Benefits the Training of Generator

In order to verify the effect of the Recurrent Residual Convolutional Neural Network on image-to-image translation, we performed a comparison between AU-GAN with plain residual convolution neural network and our proposed AU-GAN. The results are shown in Fig. 8, Fig. 9 and Fig. 10. From the results, we can find the following phenomenon: (1) AU-GAN is superior to AU-GAN with a plain residual convolution neural network in both generated images and its corresponding attention mask and content mask. Moreover, the visualization of the generator loss during training is shown in Fig. 10, in which the AU-GAN with plain residual convolution neural network and our proposed AU-GAN are demonstrated by the green curve and the red curve respectively. From them, we can find that AU-GAN is more stable and the speed of convergence is faster than

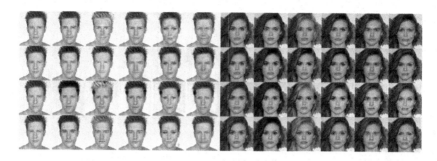

Fig. 8. The contribution of Attention Gate and a built-in attention mechanism to image-to-image translation. From top to bottom are AU-GAN without Attention Gate, AU-GAN without RRCNN, AU-GAN without a built-in attention mechanism, and AU-GAN respectively.

AU-GAN without RRCNN. I think there are two reasons for this phenomenon: one is that a residual unit helps when training deep architecture, and the other is that feature accumulation with recurrent residual convolutional layers ensures better feature representation for image-to-image tasks.

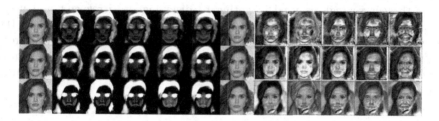

Fig. 9. The generated object mask and content mask from different models. From top to bottom are AU-GAN without Attention Gate, AU-GAN without RRCNN and AU-GAN respectively.

4.6 The Attention Gate Encourages the High-Quality Image-to-Image Synthesis

In order to verify the contribution of the Attention Gate to image-to-image translation, we performed a comparison between AU-GAN without Attention Gate and the proposed AU-GAN, which is shown in Fig. 8 and Fig. 9. The quality of the image decreased significantly when Attention Gate is removed. And the visualization of the generator loss during training is shown in Fig. 10, ??, and ?? in which the AU-GAN without Attention Gate and our proposed AU-GAN are demonstrated by the blue curve and the red curve respectively. From them, we can find that AU-GAN is more stable than AU-GAN without Attention Gate. I conjecture that the reasons why Attention Gate was very successful in the task of image-to-image translation are shown below: (1) Models trained

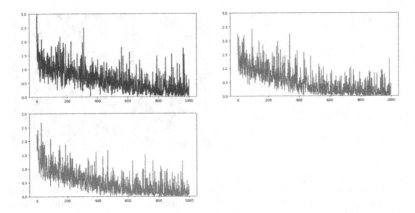

Fig. 10. The classification loss of AU-GAN without Attention Gate, without RRCNN and without AU-GAN, respectively. The horizontal ordinate is the training steps of generator G, and the ordinate scale is the classification loss of generator G. (Color figure online)

with AGs implicitly learn to suppress irrelevant regions in an input image while highlighting salient features useful for a specific task. (2) Information extracted from the coarse scale is used in gating to disambiguate irrelevant and noisy responses in skip connections. (3) AGs filter the neuron activations during the forward pass as well as during the backward pass. Gradients originating from background regions are down-weighted during the backward pass. This allows model parameters in shallower layers to be updated mostly based on spatial regions that are relevant to a given task.

5   Conclusion

In this paper, we proposed AU-GAN, which utilizes three well-known networks (the Recurrent Residual Convolutional Neural Network (RRCNN), the Attention Gate Network, as well as the built-in attention mechanism) for multi-domain image-to-image translation. Some qualitative and quantitative experiments showed that the proposed model is not only better than conventional methods but also showed more stable training. Besides, ablation experiments of AU-GAN were carried out to explore the effect of RRCNN, Attention Gate Network, and the built-in attention mechanism on the task of image-to-image translation. The task of high-resolution image-to-image translation has become a research hotspot, and we will apply the proposed model to it in the future.

References

1. Krizhevsky, A., Sutskever, I., Hinton, G.: ImageNet classification with deep convolutional neural networks. In: NIPS, vol. 25. Curran Associates Inc. (2012)

2. He, K., Zhang, X., Ren, S., Sun, J.: Deep residual learning for image recognition. In: 2016 IEEE Conference on Computer Vision and Pattern Recognition (CVPR), Las Vegas, NV, USA, pp. 770–778 (2016). https://doi.org/10.1109/CVPR.2016.90

3. Simonyan, K., Zisserman, A.: Very deep convolutional networks for large-scale image recognition. arXiv (2014)

4. Russakovsky, O., Deng, J., Su, H., Krause, J., Satheesh, S., Ma, S., et al.: ImageNet large scale visual recognition challenge. Int. J. Comput. Vis. **115**(3), 211–252 (2015). https://doi.org/10.1007/s11263-015-0816-y

5. Lin, T.-Y., et al.: Microsoft COCO: common objects in context. In: Fleet, D., Pajdla, T., Schiele, B., Tuytelaars, T. (eds.) ECCV 2014. LNCS, vol. 8693, pp. 740–755. Springer, Cham (2014). https://doi.org/10.1007/978-3-319-10602-1_48

6. Alom, M.Z., Hasan, M., Yakopcic, C., Taha, T.M., Asari, V.K.: Recurrent residual convolutional neural network based on U-Net (R2U-Net) for medical image (2018)

7. Vaswani, A., Shazeer, N., Parmar, N., Uszkoreit, J., Jones, L., Gomez, A.N., et al.: Attention is all you need. arXiv (2017)

8. Günther, S., Ruthotto, L., Schroder, J.B., Cyr, E.C., Gauger, N.R.: Layer-parallel training of deep residual neural networks. SIAM J. Math. Data Sci. **2**(1), 1–23 (2018)

9. Ming, L., Hu, X.: Recurrent convolutional neural network for object recognition. In: IEEE Conference on Computer Vision and Pattern Recognition, pp. 3367–3375. IEEE Computer Society (2015)

10. Zahangir, A.M., Mahmudul, H., Yakopcic, C., Taha, T.M., Asari, V.K.: Improved inception-residual convolutional neural network for object recognition. Neural Comput. Appl. **32**, 279–293 (2017). https://doi.org/10.1007/s00521-018-3627-6

11. Donahue, J., Hendricks, L.A., Guadarrama, S., Rohrbach, M., Venugopalan, S., Saenko, K., et al.: Long-term recurrent convolutional networks for visual recognition and description. Elsevier (2015)

12. Alom, M.Z., Hasan, M., Yakopcic, C., Taha, T.M.: Inception recurrent convolutional neural network for object recognition (2017)

13. Goodfellow, I.J., Pouget-Abadie, J., Mirza, M., Xu, B., Warde-Farley, D., Ozair, S., et al.: Generative adversarial networks (2014)

14. Isola, P., Zhu, J.Y., Zhou, T., Efros, A.A.: Image-to-image translation with conditional adversarial networks. In: IEEE Conference on Computer Vision and Pattern Recognition (2016)

15. Zhu, J.Y., Park, T., Isola, P., Efros, A.A.: Unpaired image-to-image translation using cycle-consistent adversarial networks. IEEE (2017)

16. Liang, X., Zhang, H., Lin, L., Xing, E.: Generative semantic manipulation with mask-contrasting GAN. In: Ferrari, V., Hebert, M., Sminchisescu, C., Weiss, Y. (eds.) ECCV 2018. LNCS, vol. 11217, pp. 574–590. Springer, Cham (2018). https://doi.org/10.1007/978-3-030-01261-8_34

17. Chen, X., Xu, C., Yang, X., Tao, D.: Attention-GAN for object transfiguration in wild images. In: Ferrari, V., Hebert, M., Sminchisescu, C., Weiss, Y. (eds.) ECCV 2018. LNCS, vol. 11206, pp. 167–184. Springer, Cham (2018). https://doi.org/10.1007/978-3-030-01216-8_11

18. Tang, H., Xu, D., Sebe, N., Yan, Y.: Attention-guided generative adversarial networks for unsupervised image-to-image translation. In: 2019 International Joint Conference on Neural Networks (IJCNN). IEEE (2019)

19. Zhang, H., Goodfellow, I., Metaxas, D., Odena, A.: Self-attention generative adversarial networks (2018)

20. Fu, J., Liu, J., Tian, H., Li, Y., Bao, Y., Fang, Z., et al.: Dual attention network for scene segmentation (2018)

21. Oktay, O., Schlemper, J., Folgoc, L.L., Lee, M., Heinrich, M., Misawa, K., et al.: Attention U-Net: learning where to look for the pancreas (2018)

22. Ronneberger, O., Fischer, P., Brox, T.: U-Net: convolutional networks for biomedical image segmentation. In: Navab, N., Hornegger, J., Wells, W.M., Frangi, A.F. (eds.) MICCAI 2015. LNCS, vol. 9351, pp. 234–241. Springer, Cham (2015). https://doi.org/10.1007/978-3-319-24574-4_28

23. Choi, Y., Choi, M., Kim, M., Ha, J.W., Choo, J.: StarGAN: unified generative adversarial networks for multi-domain image-to-image translation. In: 2018 IEEE/CVF Conference on Computer Vision and Pattern Recognition (CVPR). IEEE (2018)

24. Gulrajani, I., Ahmed, F., Arjovsky, M., Dumoulin, V., Courville, A.: Improved training of Wasserstein GANs (2017)

25. Sattigeri, P., Hoffman, S.C., Chenthamarakshan, V., Varshney, K.R.: Fairness GAN: generating datasets with fairness properties using a generative adversarial network. IBM J. Res. Dev. **63**(4/5), 1 (2019)

26. Karras, T., Aila, T., Laine, S., Lehtinen, J.: Progressive growing of GANs for improved quality, stability, and variation (2017)

27. Schölkopf, B.: The kernel trick for distances. MIT Press (2000)

28. Obukhov, A., Krasnyanskiy, M.: Quality assessment method for GAN based on modified metrics inception score and Fréchet inception distance. In: Silhavy, R., Silhavy, P., Prokopova, Z. (eds.) CoMeSySo 2020. AISC, vol. 1294, pp. 102–114. Springer, Cham (2020). https://doi.org/10.1007/978-3-030-63322-6_8

29. Barratt, S., Sharma, R.: A note on the inception score (2018)

Multi-robot Online Complete Coverage Based on Collaboration

Leilei Duan[1], Jianming Wang[1], and Yukuan Sun[2(✉)]

[1] School of Computer Science and Technology, Tiangong University, Tianjin 300387, China
{2031081015,wangjianming}@tiangong.edu.cn
[2] Center for Engineering Intership and Training, Tiangong University, Tianjin 300387, China
sunyukuan@tiangong.edu.cn

Abstract. Covering an unknown environment is a common task, and the multi-robot complete coverage path planning algorithm (MCPP) has received extensive attention based on its high efficiency. In the coverage task, the robot may fail or stop the task due to other reasons, and the lack of cooperative multi-robot strategy cannot deal with these situations flexibly, so this paper proposes an online multi-robot cooperative coverage method (MRCC). Considering the characteristics of path smoothness and speed constancy of the single-robot strategy Hex Decomposition Coverage Planning (HDCP), this paper extends it to multi-robot systems, and proposes a strategy that integrates communication, task sharing, and path planning based on information sharing, which can effectively deal with the problem that robots recognize each other as obstacles due to the lack of communication mechanism. When a single robot is trapped in a local area and cannot escape, the task can be shared with other team members, and the information sharing between robots can help robots plan valid path. Finally, experiments are carried out in the gazebo simulation environment, the results prove the effectiveness of our method, and the comparison experiments with the other two methods show that our method has higher efficiency.

Keywords: Multi-Robot · Complete Coverage · Collaboration

1 Introduction

Area-covering robots have been successfully applied in a variety of tasks. The goal of area-covering robots is that one or more robots traverse every obstacle-free area in the environment while avoiding obstacles, such as area cleaning [1], search and rescue missions [2], forest fire monitoring [3] and so on.

In large-scale environments, using a single robot to cover a large structure or a wide area has many disadvantages, such as time, length, robot energy, information quality and quantity. In recent years, multi-robot teams have received extensive attention. They show

This work was supported by The National Natural Science Foundation of China (62072335) and The Tianjin Science and Technology Program (19PTZWHZ00020).

high efficiency and good robustness in dealing with large-scale environmental problems. However, with the increase of team size, it also brings some challenges.

The core problem of MCPP is that individuals cooperate with each other to ensure complete and non-repetitive coverage. One approach in the task of complete area coverage is area segmentation and coverage. In a recent work [4], the author proposes an algorithm DARP, which divides the region of interest for each robot according to the initial position of the robot, and uses the Spanning Tree (STC) algorithm for each robot in the allocated region [5]. The premise of this algorithm is to understand the entire environment. However, in many scenarios, there is usually no prior knowledge as a reference. It is necessary to collect surrounding data in real time through onboard sensors and make corresponding path planning. Whether robots can obtain real-time information about the surrounding environment and communicate with each other affects the completion efficiency of the entire task.

The previously described work does not pay much attention to the cooperation between robots. This paper proposes an effective online coverage algorithm MRCC, which enables the robots to understand each other's task completion in real time, and can apply the information that has been explored to the path choose aspect. A single robot may fail in its own area or there are obstacles that cannot be bypassed. For this situation where it cannot continue to explore, our proposed method can share its unfinished tasks with other robots, improving the flexibility of tasks. Considering that in an unknown environment, the time to complete the entire space exploration is unpredictable, for the robot energy problem, it is crucial to complete the task in a short time, Therefore, for each robot coverage task, we adopt the single robot coverage strategy Hex Decomposition Coverage Planning (HDCP) proposed by [6]. This algorithm MRCC has been verified by the simulation environment Gazebo. The rest of this article is organized as follows. Section 2 briefly reviews existing MCPP algorithms, Sect. 3 elaborates the details of the MCPP problem and the MRCC algorithm, the results are discussed in Sect. 4, and the paper ends in Sect. 5 with recommendations for future work.

2   Related Works

Various factors affect the results of MCPP, including task allocation methods, information sharing methods (centralized, distributed), and path generation methods. Compared with the single-robot coverage path planning algorithm (SCPP), MCPP can guide multi-robots to complete tasks in parallel, thereby reducing the time consumption of the task is improved, and the execution efficiency of the task is improved.

Many multi-robot CPP methods proposed in literature are extensions of traditional single-robot CPP methods. In order to solve the problem of redundancy and collisions between robots, The methods in [7, 8] performed cell decomposition, deploying each robot to explore different sub-regions. The work in [9, 10] carried out grid-based decomposition of 2D regions of interest. The decomposed areas are used to plan paths and assign areas to robot teams. Based on the single-robot STC algorithm, Hazon et al. proposed the multi-robot STC methods (MSTC) to solve the multi-robot coverage path planning CPP problem [11, 12].

Coverage exploration methods are mainly divided into model-based and non-model-based methods. Model-based methods rely on the reference model or structure of the environment and are known to the robot. Robots have a full understanding of the region and have been proved to be NP-hard [13]. Non-model-based approaches plan and explore without prior knowledge of structure or environment [14, 15]. On the basis of this classification, the viewpoint is generated and the search space is formed. In the large-scale coverage problem where the location and number of required viewpoints cannot be determined in advance, the model-based approach is the primary choice. In [16], the author firstly maps the entire environment, then transforms the MCPP problem into a multi-traveling salesman problems (MTSP) solution problem, and finally genetic algorithm (GA) [17] algorithm is responsible for assigning each region to the robot. The whole process focuses on the effective assignment of tasks, but the trajectory generated by it is not smooth enough, there are many overlaps of trajectory, and the process of drawing consumes time.

Meanwhile, sampling-based exploration methods have been widely applied in practical exploration systems, mainly including RRT family [18, 19]. Although the sample-based method has many improvements in sample efficiency and computational overhead, it still has a large overlap rate. In view of the problem that artificial potential field (APF) [20] is prone to fall into local minimum and is difficult to escape, [21] proposed multi-robot multi-objective potential field (MMPF), which can solve the root cause of track overlap in sample-based methods, but the position recognition module has a great impact on the calculation cost and map quality. And it is difficult to find the target point in the large empty scene.

An algorithm that can generate smooth paths is particularly important for improving efficiency. The representative geometry-based Boustrophedon decomposition algorithm [22] generates paths with many turns, which consumes a lot of energy for the robot. Based on Dubins vehicle constraints, a fast online exploration method [6] based on hexagonal decomposition is designed, which can produce smooth paths and ensure constant travel speed. However, a single robot has limitations in covering tasks in a scene full of complex obstacles.

Our contribution: (1) Based on the feature that SCPP [6] can explore the unknown environment at a constant speed and path smoothness, this paper extends it to the multi-robot system, and the whole system does not depend on the prior information of the environment. (2) Considering the importance of coordination for multi-robot coverage task, a strategy combining communication, task sharing and path planning based on information sharing is proposed to improve the flexibility of task and the efficiency of area coverage.

3 Proposed Methodology

The traditional MCPP method mainly consists of two steps: first, the entire scene is divided into the same number as the robot. Second, each area is assigned to the robot to perform the coverage task. This approach generally requires an understanding of the environment. However, when performing many tasks, the entire environment cannot be perceived in advance, so it is difficult to implement the region segmentation algorithm in an unknown environment. In general, the size of the scene can be estimated. We can discretize the entire scene into cells. As the task progresses, we can not only understand the area coverage, but also dynamically adjust the task. We allocate the area equally to each robot to ensure the balance of tasks. Figure 1 shows our overall framework outlining the workflow, which is explained in detail in later sections.

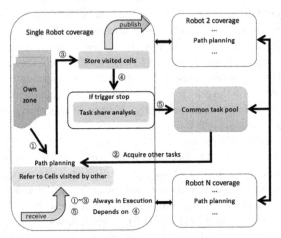

Fig. 1. Overview of the MRCC framework. Every robot contains a task decision mechanism and communicates with other robots in real time.

3.1 Multi-robot Coverage Path Planning Problem

Suppose $V = \{V_1, V_2, ..., V_N\}$ is a team composed of $N \in N^+$ robots, R representing the entire scene, each robot is responsible for an area R_i, and discretizes it into equal cells $R = \{x, y : x \in [1, \text{rows}], y \in [1, \text{cols}]\}$, where rows, cols are the number of rows and columns after discretization of the terrain to be covered. Obviously, the number of all terrain cells is given by $n = \text{rows} \times \text{cols}$. The coverage task of the entire area can be described as $\{L_1 \cup L_2 \cup, ..., L_N\} \subseteq R$, L_i is the set of cells where the coverage of the area that the robot v_i is responsible for is completed at a certain moment.

In this paper, single robot online CPP algorithm is used for coverage control of each task. And each robot maintains a decision-making mechanism inside, will take measures according to the situation.

There are many options for single-robot coverage algorithms [4]. Divide the area for each robot based on the initial position, and use the STC [5] coverage algorithm in each area. One disadvantage of STC is that the generated path has sharp bends. It is said that in these sharp corners, it is necessary to decelerate and then accelerate, which will produce a lot of energy consumption. Kan, Xinyue et al. [6] have proposed a good solution for unknown environments full of obstacles, which can generate smooth trajectories and guarantee constant speed, so apply it to our work.

3.2 Communication Mechanism

In a single robot system, robots can regard any object in the environment as an obstacle, but in a multi-robot system without information sharing, the problem of identifying each other as an obstacle needs to be solved urgently. As shown in Fig. 2, when there are large obstacles in both R2 and R3 areas, the robots need to explore a path to the unfinished area after completing some tasks in the area. In this exploration process, if the robots approach each other, they will be identified as obstacles, resulting in low area coverage.

First, any robot v_i maintains a local map, with the starting point set as the origin, and the corresponding entire scene as a global map. Members of the robot team share their operating areas with other members at any time, including their locations. In order to make full use of each other's information, mapping between maps is necessary.

$$G_i = f_{loc}(L_i) \ i = 1, 2, ..., N \tag{1}$$

f_{loc} is the mapping function from local coordinates to global coordinates, and G_i is the transformed global coordinate information.

Secondly, we use a distributed communication structure to maintain communication between any two robots, and the failure of any robot will not affect the communication between other robots, which enhances the robustness of the system.

3.3 Task Share Analysis

Assuming that the energy of the robot can meet the task requirements, the robot can generate a feasible solution to the coverage problem in limited time under the ideal state without any obstacles. In the unknown environment, the robot starts to explore with its own position as the starting point, and it may be trapped in the local area due to obstacles, thus resulting in coverage gap. Some researchers have proposed the idea of game theory [23]. When a robot fails, the robots around it will weigh the task value of its current task and that of the failing robot. If the task value of the failing robot is high, it will give up its own task and prioritize the task with high value. The downside of this approach is that the value of an area in an unknown environment is difficult to estimate, and travel costs are sacrificed when returning to one's mission.

In DARP [4], each robot is only responsible for its own regional tasks, and there is no task adjustment strategy. When a robot fails to complete the regional tasks due to other reasons, other robots cannot assist the robot, which will lead to incomplete coverage. So, we came up with a solution. After a robot v_i stops exploring, it checks the task quantity L_i completed at the current moment. If it is greater than the set threshold $K = R_i - L_i$, it indicates that it has broken down or is trapped in a certain area. The process of sharing tasks with other robots includes:

$$R = \{L_i(V_i)\} \cup \{L_j(V_j), \ v_j \in (V - v_i) : (L_j = R_j)\} \tag{2}$$

This includes the current tasks of all robots. This condition $L_j = R_j$ ensures that other robots finish their tasks first, and that any area is equally high in priority, and that the rest of the team can choose to cover the remaining common areas at the same time.

The algorithm process is given by Algorithm 1, where represents the number of tasks assigned to the robot and is a common task set.

Algorithm 1 Task shared Process

Require: initialize cell set U_i tasks sharing set C

1: **while** $U_i \neq \emptyset$ **do**
2:     Radar scans the surroundings
3:     **repeat**
4:         **if** *cell* is obstacle **then**
5:             Check whether it is another robot
6:             **if** it is the other robot **then**
7:                 pass
8:             **else**
9:                 remove it from U_i
10:             **end if**
11:         **else**
12:             Access and add *cell* $\rightarrow L_i$,then delete it from U_i
13:         **end if**
14:     **until** trigger stop sign
15: **end while**
16: **if** $U_i - L_i \neq \emptyset$ **then**
17:     $(U_i - L_i) \rightarrow C$
18:     Repeat steps 1-14 and replace U_i with C
19: **end if**

3.4 Coverage Path Pre-planning

Members of the robot team trigger stop signs on two occasions while completing the area coverage task. (1) Assigned tasks and common tasks have been completed, and (2) stopped for other reasons. In the second case, coverage gap is usually generated because there are obstacles around the robot and it is trapped in place due to physical limitations and cannot continue to work, or there is an impassable obstacle in front of the robot and the boundary of the region is on both sides, as shown in Fig. 2 for R2.

We put forward the task of sharing strategy can solve the problems of the former, while for the latter, we put forward the path solution, in the process of the planning of a robot $v_i \in V$ from cell $\left(x_i, \ y_i\right)$ to $\left(x_j, \ y_j\right)$ need a path, the robot in the unfinished area select a cell $(x, \ y) \in (R_i - L_i)$ as target, the generated path to satisfy the following conditions:

$$p = \left(\begin{matrix} \text{info} \\ \text{max} \end{matrix}\right) \left(\begin{matrix} \text{dis} \\ \text{min} \end{matrix} \left\{ \left(x_i, \ y_i\right) \rightarrow \left(x_j, \ y_j\right) \right\}\right) \tag{3}$$

dis stands for the shortest distance between two cells, and the role of info is the current area coverage information L_i of other robots that the robot can refer to when pre-planning the path.

In the absence of communication, robots walk while exploring. If they can plan in advance by referring to the coverage information of other robots before moving, they can avoid infeasible paths in advance, which will shorten the time to reach the target point to some extent.

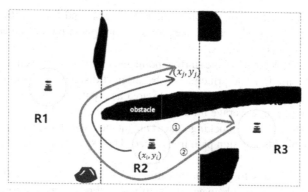

Fig. 2. Each robot is responsible for an area, and there are obstacles of different sizes in the area. Some obstacles may block the normal exploration of the robot.

Algorithm 2 Coverage path pre-planning

1: **procedure** :(pre-planning)
2:    initialize a cell list $Candidates$ and dict $past\_candidates$
3:    set $specific\_way\_flag=False$ and $uncertain\_way\_flag=False$
4:    **while** $Candidates \neq \emptyset$ **do**
5:        Pop up a $candidate$ from $Candidates$ that is closest to the target
6:        **for** $cell$ in $neighbors$ **do**
7:            **if** If the $cell$ state is unknown and appears for the first time **then**
8:                record the $cell$ and modify $uncertain\_way\_flag$
9:            **else**
10:                **if** $cell$ state is known **then**
11:                    **if** $cell$ not in $past\_candidates$ or new_cost(cell, target)↓ **then**
12:                        update $cell \rightarrow candidates$, $cell \rightarrow past\_candidates$
13:                    **end if**
14:                **else**
15:                    continue
16:                **end if**
17:            **end if**
18:        **end for**
19:        Modify flags when the target area is reached or $Candidates = \emptyset$
20:    **end while**
21: **end procedure**

Given the unknown environment, the paths generated using some path planning algorithms cannot be determined for their feasibility (it is uncertain whether the cells in the path are occupied by obstacles). Figure 2 shows the process of generating a path based on the shortest distance. Robot in R2 chooses the red route ① when it goes to the target point $\left(x_j, y_j\right)$. . During the process of exploring while walking, it is found that this shortest path is infeasible (There are impassable obstacles in R3), and then the robot has to choose other paths such as the red route ②, and this backtracking process increases the travel cost. Therefore, we propose a pre-planning mechanism. Algorithm 2 gives the pre-planning process.

After the robot determines the starting point and target point, we choose certain cells to form a path based on the principle of the shortest distance. For example, the robot in R2 selects the cells belonging to the R3 area as a path component. However, after referring to the current coverage information (cell status) of R3, we learned in advance that the red route ① is the blocking state. Our method will turn to other directions to find paths to avoid the invasion path in advance. The experimental results are shown in Fig. 5.

There are two ways to generate feasible paths (blue paths in the figure): (1) If a definite path can be formed between the start and end positions, the robot will choose this path to reach its unfinished area through other areas. We prefer the path that can be determined, because this guarantees the feasibility of the path. The characteristics of the cells that make up this determined path are that it has been determined by any robot to be not an obstacle, that is, the state is known. (2) If a definite path cannot be generated, the robot in R2 avoids the red path and starts to explore while walking from the first cell whose state is unknown. Algorithm 3 clarifies how to make decisions on the next action of the robot according to the information fed back by Algorithm 2.

Algorithm 3 Determine what the robot will do next based on Alg.2

1: **procedure** :(path)
2:    **if** *specific_way_flag* **then**
3:        return Path
4:    **else**
5:        **if** *uncertain_way_flag* **then**
6:            start exploring from the recorded *cell*
7:        **else**
8:            stop exploration
9:        **end if**
10:    **end if**
11: **end procedure**

---

4   Experiments and Discussion

The MRCC algorithm was validated on Gazebo robot simulation platform using an I9-19850K computer with a 3.60 GHz CPU and 32 GB RAM. We provide multiple 20 × 20 m scenarios with different internal obstacles, which are similar to a more realistic outdoor environment. The robot model we applied is TurtleBot3, which is equipped with 2D radar and has a scanning range of 3.5 m. The hexagon element has side length r = 1 m. The robot moves at a constant speed of 1 m/s. The coverage time of a robot is the total time from the start of the robot to the completion of the assigned area. For multiple robots, the exploration time is the average of all robot times.

This paper sets up experiments for the problems described in the previous chapters as follows: (1) Solve the problem that robot team members recognize each other as obstacles. (2) The robot cannot escape the local area, and the task is shared with other members of the team. (3) The path pre-planning problem. (4) Compare our algorithm with RRT and MMPF in time.

We first applied three robots to conduct experiments in Fig. 3(a) scene, which was vertically divided into three areas, and each robot performed area coverage tasks in its own responsible area. Red cells represented obstacles, green cells were visited, and other color lines represented trajectories of different robots. Other experiments in this paper also adopt the same representation method. We can see from Fig. 3(b) that obstacles block the exploration of the upper part of Robot2 and Robot3. In the process of exploring other paths, since there is no information communication between robot team members, Robot3 will regard the cell where Robot2 is located as an obstacle (the part marked in red in Fig. 3(b), thus creating a coverage gap. Figure 3(c) show the effectiveness of our method.

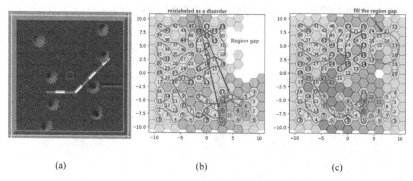

Fig. 3. Experimental results of communication mechanism (Color figure online)

In Fig. 4(a), we simulate a new scene, the middle part of the scene is completely blocked by obstacles, and the robot2 is used as the research object to test the situation of other members sharing the work after it is trapped in the local area. Figure 4(b) is the result when there is no task sharing mechanism. Robot2 first performs the coverage task in this area. After there are no selectable cells, it searches for a path and finds that it cannot cross the obstacle, and the task is forced to terminate, thus resulting in cover the gap. The results in Fig. 4(c) show that after robot2 cannot escape the local area, it sets its unfinished task as a public task. Robot1 and robot3 detect that there is remaining work in the area, and will fill the coverage gap. In the case of completing the task as the first principle, due to the existence of this mechanism, high regional coverage will be guaranteed.

Pre-planning paths based on information from other robotics team members can improve efficiency. Figure 5(b) records the process of robot2 going to the remaining area after completing the local area task. When there is no pre-planning strategy, robot2 starts from cell 10 and explores the path while walking. When the path is found to be

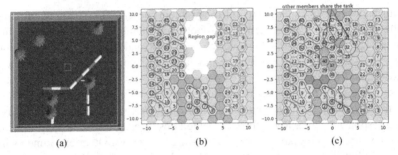

Fig. 4. Task sharing experiment comparison (Color figure online)

invalid (the red marked part in Fig. 5(b)), it needs to go back a certain distance and re-explore (the blue line marked part). Figure 5(c) is the result containing the pre-planning strategy. This infeasible solution can be avoided in advance by referring to the map information obtained by Robot3. Figure 5(d) shows that our strategy not only shorted coverage time of Robot2, but also improved overall efficiency.

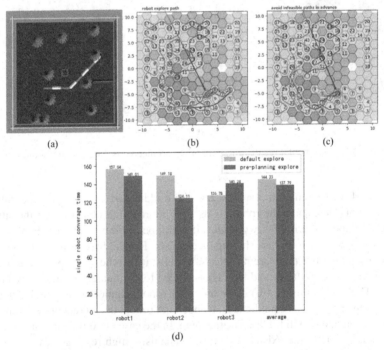

Fig. 5. Experimental results of path pre-planning (Color figure online)

Figure 6 shows our comparison results with RRT and MMPF. The algorithm RRT continuously samples within the sensor range and selects an optimal sampling point as the exploration target. In the exploration process, it often lingers between the target points, increasing the travel cost. As for the defects of RRT, MMPF tends to send robots to different frontier clusters to explore multiple places at the same time. The robot will continue to explore one side without turning back until it finishes exploring this side. However, in a slightly empty scene, the robot will be unable to find the target point, causing the robot to stay in place. The goal of multi-robot cooperation is to share the tasks in the whole region and improve the overall efficiency. However, in MMPF, only when there are similar descriptors between robots, they will add their own map to each other's map. If there are no similar descriptors, the whole task is still completed by each robot alone, and the cooperation between robots can only depend on the occurrence of map merger.

We respectively applied a single robot and two robots to conduct comparative experiments in a scene with a size close to 400 square meters. After several experiments, it can be seen in Fig. 6 that the time taken by a single RRT and MMPF robot to complete tasks is not much different from the time taken by the two robots to complete tasks together. This is because MMPF did not allocate the whole area in advance, and the robots could not properly exchange their completed tasks when they met, resulting in a large number of repetitive tasks, which was not sufficient in the performance of task coordination. And the process of selecting target point is accompanied by deceleration and acceleration, which reduces the efficiency. In our algorithm, at the beginning of the task, each robot has its own unique area, tasks are carried out in parallel, but the team is not completely independent, so it can make timely decisions according to some situations encountered. In terms of decision-making, the location of obstacles can be determined with the help of radar information, which can be quickly marked on the map, and target points can be quickly selected. The dynamic feasibility of Dubins vehicle is taken into account in the way of movement, and the vehicle maintains a constant speed, so it has obvious advantages in terms of time.

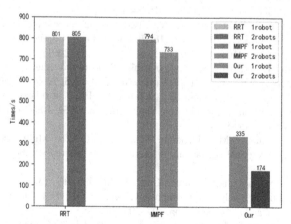

Fig. 6. Comparison with RRT and MMPF

5 Conclusion

This paper proposes a multi-robot collaborative online coverage algorithm suitable for unknown environments, analyzes the problems that affect the coverage rate and coverage efficiency of the traditional MCPP method based on SCPP extension in the coverage task, and provides information communication, task sharing, path pre-planning measures improve the flexibility of tasks and enhance the robustness of multi-robot teams. Studies that can both take into account dynamic feasibility in some form (Dubins vehicles in this case) and ensure full coverage are limited, our MRCC applies HDCP algorithms that take this into account to multi-robot teams, the results show good performance in multi-robot teams. There are also some shortcomings in this paper. The shared tasks lack an allocation mechanism, and team members do not measure how many tasks they should undertake according to their individual task load. Our future work will focus on efficient collaboration mechanisms.

References

1. Janchiv, A., Batsaikhan, D., Kim, B., Lee, W.G., Lee, S.: Time-efficient and complete coverage path planning based on flow networks for multi-robots. Int. J. Control Autom. Syst. **11**, 369–376 (2013). https://doi.org/10.1007/s12555-011-0184-5
2. Xu, A., Viriyasuthee, C., Rekleitis, I.: Efficient complete coverage of a known arbitrary environment with applications to aerial operations. Auton. Robot. **36**(4), 365–381 (2013). https://doi.org/10.1007/s10514-013-9364-x
3. Merino, L., Caballero, F., Dios, J.R., Maza, I., Ollero, A.: An unmanned aircraft system for automatic forest fire monitoring and measurement. J. Intell. Robot. Syst. **65**, 533–548 (2012). https://doi.org/10.1007/s10846-011-9560-x
4. Kapoutsis, A.C., Chatzichristofis, S.A., Kosmatopoulos, E.B.: DARP: divide areas algorithm for optimal multi-robot coverage path planning. J. Intell. Robot. Syst. **86**(3–4), 663–680 (2017). https://doi.org/10.1007/s10846-016-0461-x
5. Gabriely, Y., Rimon, E.D.: Spanning-tree based coverage of continuous areas by a mobile robot. In: Proceedings 2001 ICRA. IEEE International Conference on Robotics and Automation (Cat. No. 01CH37164), vol. 2, pp. 1927–1933 (2001)
6. Kan, X., Teng, H., Karydis, K.: Online exploration and coverage planning in unknown obstacle-cluttered environments. IEEE Robot. Autom. Lett. **5**, 5969–5976 (2020)
7. Rekleitis, I.M., New, A.P., Rankin, E.S., Choset, H.: Efficient boustrophedon multi-robot coverage: an algorithmic approach. Ann. Math. Artif. Intell. **52**, 109–142 (2009). https://doi.org/10.1007/s10472-009-9120-2
8. Azpúrua, H., Freitas, G.M., Macharet, D.G., Campos, M.F.: Multi-robot coverage path planning using hexagonal segmentation for geophysical surveys. Robotica **36**, 1144–1166 (2018)
9. Gautam, A., Jatavallabhula, K., Kumar, G., Ram, S.P., Jha, B., Mohan, S.: Cluster, allocate, cover: an efficient approach for multi-robot coverage. In: 2015 IEEE International Conference on Systems, Man, and Cybernetics, pp. 197–203 (2015)
10. Modares, J., Ghanei, F., Mastronarde, N., Dantu, K.: UB-ANC planner: energy efficient coverage path planning with multiple drones. In: 2017 IEEE International Conference on Robotics and Automation (ICRA), pp. 6182–6189 (2017)
11. Hazon, N., Kaminka, G.A.: Redundancy, efficiency and robustness in multi-robot coverage. In: Proceedings of the 2005 IEEE International Conference on Robotics and Automation, pp. 735–741 (2005)

12. Hazon, N., Kaminka, G.A.: On redundancy, efficiency, and robustness in coverage for multiple robots. Robot. Auton. Syst. **56**, 1102–1114 (2008)
13. Arkin, E.M., Fekete, S.P., Mitchell, J.B.: Approximation algorithms for lawn mowing and milling. Comput. Geom. **17**, 25–50 (2000)
14. Scott, W.R.: Model-based view planning. Mach. Vis. Appl. **20**, 47–69 (2007). https://doi.org/10.1007/s00138-007-0110-2
15. Scott, W.R., Roth, G., Rivest, J.: View planning for automated three-dimensional object reconstruction and inspection. ACM Comput. Surv. **35**, 64–96 (2003)
16. Sun, R., Tang, C., Zheng, J., Zhou, Y., Yu, S.: Multi-robot path planning for complete coverage with genetic algorithms. In: Yu, H., Liu, J., Liu, L., Ju, Z., Liu, Y., Zhou, D. (eds.) ICIRA 2019. LNCS (LNAI), vol. 11744, pp. 349–361. Springer, Cham (2019). https://doi.org/10.1007/978-3-030-27541-9_29
17. Reeves, C.R., Rowe, J.E.: Genetic Algorithms: Principles and Perspectives: A Guide to Ga Theory (2002)
18. LaValle, S.M.: Rapidly-exploring random trees: a new tool for path planning. The annual research report (1998)
19. Umari, H., Mukhopadhyay, S.: Autonomous robotic exploration based on multiple rapidly-exploring randomized trees. In: 2017 IEEE/RSJ International Conference on Intelligent Robots and Systems (IROS), pp. 1396–1402 (2017)
20. Warren, C.W.: Global path planning using artificial potential fields. In: Proceedings 1989 International Conference on Robotics and Automation, vol. 1, pp. 316–321 (1989)
21. Yu, J., et al.: SMMR-explore: submap-based multi-robot exploration system with multi-robot multi-target potential field exploration method. In: 2021 IEEE International Conference on Robotics and Automation (ICRA), pp. 8779–8785 (2021)
22. Choset, H., Pignon, P.: Coverage path planning: the boustrophedon cellular decomposition. In: Zelinsky, A. (ed.) Field and Service Robotics. Springer, London (1998). https://doi.org/10.1007/978-1-4471-1273-0_32
23. Song, J., Gupta, S.: CARE: cooperative autonomy for resilience and efficiency of robot teams for complete coverage of unknown environments under robot failures. Auton. Robot. **44**(3–4), 647–671 (2019). https://doi.org/10.1007/s10514-019-09870-3

A Long Short-Term Urban Air Quality Prediction Model Based on Spatiotemporal Merged GLU and GCN

Wenjing Xu[1,2(✉)], Jie Hao[1,2], and Shifang Lu[3]

[1] College of Computer Science and Technology, Nanjing University of Aeronautics
and Astronautics, Nanjing, China
{xu.wenjing,haojie}@nuaa.edu.cn
[2] Collaborative Innovation Center of Novel Software Technology
and Industrialization, Nanjing, China
[3] Cetc Information Technology System Co., Ltd., Nanjing, China

Abstract. Long-term prediction of PM2.5 concentration plays a great guiding role in strengthening urban pollution prevention and control. In recent years, some short-term prediction models based on RNN and its variants LSTM and GRU have been well studied, whereas the recurrent structure is difficult to mine long term dependence of dynamic air quality data. In this paper, we process the spatial correlation among air quality monitoring stations using a graph, and analyze urban heterogeneous data including surrounding point of interest (POI) data to capture the spatial dependency. Finally, we propose a novel long short-term prediction model that fuses gated linear units (GLU) and graph convolutional neural network (GCN). Benefiting from its complete convolutional structures, the proposed model can quickly capture the long term spatiotemporal dependence of the stations and greatly improves the reasoning speed. We have conducted extensive experiments based on city data to verify the validity of the proposed model, and compared our model with 6 baseline models. The results show that our model reduces the short and long term prediction errors by 5.9% and 16.8%, respectively.

Keywords: Air quality prediction · Spatiotemporal dependence · Long-term prediction · Big data

1 Introduction

In recent years, air pollution has received widespread attentions. Fine particulate matter (PM2.5) is the air pollutant that poses the greatest risk to health globally. Chronic exposure to PM2.5 considerably increases the risk of respiratory

This work is supported in part by the National Key R&D Program of China under Grant 2019YFB2102000, and in part by the Collaborative Innovation Center of Novel Software Technology and Industrialization.

and cardiovascular diseases in particular [18]. Therefore, long-term prediction of PM2.5 concentrations is a key concern for the health of urban residents. With the construction and development of smart cities, more and more micro air quality monitoring stations are being deployed in cities, and increasing amount of urban big data for PM2.5 forecasting are available. In addition, the air quality of a location is not only related to the location but also to its point of interest attribute around the location, for example, the air quality around a park is generally better than the air quality near a factory [8]. However, the involving urban data are generally characterised by multi-source heterogeneity, dynamic variability and spatiotemporal correlation, making the long short-term prediction of urban air quality a challenging task.

The research on air quality prediction has achieved great progress, especially on short-term prediction. However, the research on long-term air quality prediction still needs to be enriched. Existing short-term prediction methods can be summarised into three categories: statistical models incorporating domain knowledge, data driven deep learning models that only consider temporal or spatial correlation, and hybrid deep learning models which consider both spatiotemporal correlation. Firstly, statistical models, such as regression analysis models and cluster analysis models, require complex domain knowledge, and are difficult to adapt to the dynamics. Secondly, data driven deep learning models, such as DNNs [17], have powerful feature representation capability while ignoring the correlations among monitoring stations. In order to capture the inter-station spatial correlations, GCN and CNN are introduced in some hybrid deep learning models [4,10,16]. In most existing hybrid models, CNN have been used to extract the spatial dependence, which is unsatisfactory in tackling irregular spatial dependence caused by the unevenly distributed stations [3,8]. In these models, RNNs and their variants LSTM are usually used to extract temporal dependence [3,5,12,14]. However, the recurrent structure of RNN processes sequence data one by one, which cannot realize parallel operation and runs slowly. In addition, the extractions of spatial and temporal dependence are separated in these hybrid models, greatly increasing the data noise and difficulty to capture long-term dependence.

To address the problems mentioned above, we take into account the spatial-temporal information among air quality stations and POI attribute around the stations, and propose a long short-term urban air quality prediction model based on spatiotemporal merged GLU and GCN. Specifically, convolutional neural network with a gating mechanism is used to extract the long temporal dependence to avoid the long recurrent structure of RNNs. To better extract the inter-station spatial dependence, we use the graph structure to model the unevenly distributed stations and consider the POI data around the stations to capture the impact of ambient environment. Meanwhile, spatial and temporal feature extraction blocks are fused into one structure that can deeply capture the long-term spatiotemporal dependence and reduce prediction errors. The main contributions in this paper can be summarised as follows:

(1) We exploit the urban heterogeneous data with strong spatial correlation (i.e., POI data) and stations' monitoring data as our model input and we

fuse GLU and GCN in one structure to extract spatiotemporal dependence simultaneously, which can deeply mine the long-term dependence of the historical data.

(2) We conduct extensive experiments and compare our proposed method with 6 baseline models. The experiment results show that for short-term prediction our method reduces the prediction errors approximately by 5.9% and 16.8% for long-term prediction approximately.

The rest of the paper is organized as follows: Sect. 2 reviews related research about air quality prediction. Section 3 formulates the air quality prediction problem in detail. Section 4 elaborates the proposed method. In Sect. 5, we evaluate the proposed method based on a real world city wide dataset. We conclude the paper in Sect. 6.

2   Related Work

The research of air quality prediction has a long history for its potential practical value. In the very beginning, most widely used methods were statistical models based on domain knowledge [8] and traditional machine learning methods such as support vector regression (SVR) [11], autoregressive integrated moving average (ARIMA) [6], and back propagation neural networks (BP). Bai et al. [2] decomposes air quality data into different scales and use a BPNN incorporating wavelet techniques to achieve short-term prediction. For small samples, statistical methods are usually no less effective than deep learning algorithms and are more interpretable. However, they require a large number of numerical calculations, which can not achieve satisfactory accuracy in terms of computational speed and inference, and require domain expertise.

Deep learning has been proved to be an effective technique for air quality modelling and prediction, and more and more air quality prediction studies based on deep learning have emerged. Yu et al. [18] establish a U-air model to make real-time prediction. However, due to the limited and unevenly deployed monitoring stations, the prediction errors are rather high. Yi et al. [17] propose a DNN based air quality prediction method that introduces domain knowledge to DNN. But the full connected operation treats spatial correlation among stations equally and does not capture the varied spatial correlation very well. The convolution operation in CNN is only suitable for handling regular data, such as images, text, and not suitable for spatiotemporal data. Therefore, GCN [9], specifically handles graph-structured data, such as social networks, is proposed for spatial data processing. Air quality data is spatio-temporally correlated, so with the proposal of GCN, more and more air quality related studies are considering the use of GCNs to capture the spatial correlation of the stations [16].

RNN [12,13], and its variants GRU [7] and LSTM [3], are now widely used for sequential data processing due to their gating mechanism. The gating mechanism is flexible enough to retain the strongly correlated parts of sequential data and discard the weakly correlated parts during the training process. Xu et al. [3] propose a multi-task model based on LSTM to predict PM2.5 within a city. However, the RNNs still treat each monitoring station as independent and can not exploit the spatiotemporal relationships among all the stations. Therefore, different combinations such as GCNs, GNN&RNN, have drawn increasing attentions to capture the spatiotemporal correlation. We refer to these models as hybrid models. Wang et al. [15] use GNN&GRU, combining domain knowledge of pollutant dispersion to address the intra-city interdependent effects. Although hybrid models are good at improving prediction accuracy, its temporal and spatial extractor are separate, in which the data transfer along the separate extractors increases the noise, and can not extract spatiotemporal dependence sufficiently. Thus, they are inadequate for long-term prediction.

In this paper, considering the inter-station spatiotemporal correlation and also associated POI data, we propose a complete convolutional model that fuses the temporal and spatial dependence in one block. It extracts the spatiotemporal dependence exactly and faster to make long short-term prediction.

3   Problem Formulation

In this paper, our goal is to predict the air quality in a certain period of time based on historical data obtained from the air quality monitoring stations deployed in urban and the POI data surrounding the stations. To better explore the inter-station spatial dependence, we consider that the stations form a complete graph $G = \{N, E\}$, where N represents all the air monitoring stations in a city with size $|N| = n$, E represents the edge set. As shown in Fig. 3, every station has a $f(f \in \mathbb{N})$ dimensional feature vector, and all the air quality stations are integrated into one graph structure. At time t, the feature vector of station i is denoted as $x_i^t \in \mathbb{R}^{f \times 1}$. Considering the functionality of the region where the monitor station is deployed influences the air quality highly, the feature vector includes the measured air quality data and also POI data. Thus $X^t = \{x_1^t, x_2^t, \ldots, x_n^t\}$ is the snapshot of all the stations' feature values at time t. Given current time instance T and a historical time window m (in our model $m = 72$ h), we define the feature matrix of G as $X = \{X^{T-m+1}, X^{T-m+2}, \ldots, X^T\}$. Therefore, the problem of the air quality prediction can be modeled as learning the mapping function F on G and X to predict the PM2.5 concentration with multi-horizon, i.e. from time $T + 1$ to $T + \tau$, where $\tau \in \mathbb{N}$ is the largest prediction time horizon. It can be formulated as

$$F(X, G) = \left\{\hat{Y}^{T+1}, \hat{Y}^{T+2}, \ldots, \hat{Y}^{T+\tau}\right\}, \tag{1}$$

where $\hat{Y}^t = \{\hat{y_1}^t, \hat{y_2}^t, \ldots, \hat{y_n}^t\}$ and $\hat{y_i}^t$ is the predicted value of station i at time t. We can see that from Eq. (1), the proposed model can yield long short-term prediction results spanning from time $T + 1$ to $T + \tau$. In the next section, we'll describe how we design the prediction model F.

4   Spatiotemporal Merged Model

In order to deeply capture the spatiotemporal dependence from urban heterogeneous data, we propose a spatiotemporal merged model based on GLU and GCN. As illustrated in Fig. 1, our model consists of stacking spatiotemporal merged blocks, each block integrates a GLU and a two-layer GCN, which are noted as temporal dependence extractor and spatial dependence extractor, respectively. In the model, with the feature matrix X as the input, firstly GLU is used to extract temporal dependence of each station. Next, the features yielded by GLU operation is feed into the two-layer GCN to extract the spatial dependence according to the constructed graph G. In order to fuse temporal dependence extractor and spatial dependence extractor in one spatiotemporal block, we use *in*, *out* channels for the number of input channels and outputs for extractor, respectively. In the first two spatiotemporal blocks, $in = out = f$. We stack the spatiotemporal merged block to extract the spatiotemporal dependence deeply and coherently. After extracting the spatiotemporal dependence by stacking layers, a full connected layer is used to obtain the prediction results. In this section we will describe the temporal and spatial dependence extractors in details.

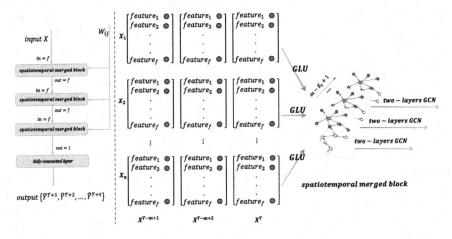

Fig. 1. The structure of the proposed spatiotemporal merged model, by stacking the combination of GLU and a two-layer GCN.

4.1   Temporal Dependence Extractor

RNNs and LSTMs [12,13] are widely used for processing sequential data through their chained structure. However, the chain-loop structure prevents parallel computation, and thus is not efficient to process long term historical air quality data.

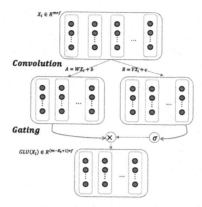

Fig. 2. The GLU structure is to obtain temporal dependence. $X_i \in R^{m \times f}$ is the i-th station's feature vector under m historical time periods, the length of the sequence after temporal dependence extraction is shortened to $m - K_t + 1$.

Incorporating gating mechanisms into CNN, GLU is able to compute in parallel on large-scale historical data and extract longer-term time-dependent features more quickly. As it provides a linear path to solve the gradient explosion and converges and is faster than RNNs, we select GLU as the temporal dependence extractor.

In the proposed temporal dependence extractor as shown in Fig. 2, for station i, firstly, its historical feature values $X_i = \{X_i^{T-m+1}, X_i^{T-m+2}, \ldots, X_i^T\}$ are position embedding to mark the original position in the sequential data, then the data will undergo two different one-dimensional convolution. The convolution results will be denoted as A, B respectively, where $A \in \mathbb{R}^{(m-K_t+1) \times out}$, $B \in \mathbb{R}^{(m-K_t+1) \times out}$. in, out are the number of input, output channels. Initially, in, out are the number of features. K_t is time convolution kernel size of the temporal dependence extractor, in our model we choose $K_t = 3$. Among these, A is used as the temporal dependence extractor for the current series data and a nonlinear transformation is performed on B by using an activation function to pass useful dependence information through the gating mechanism. Thus the temporal dependence extractor can be defined as

$$H_l(X_i) = (WX_i + b) \otimes \sigma(VX_i + c) \tag{2}$$

where $W \in \mathbb{R}^{K_t \times in \times out}$, $V \in \mathbb{R}^{K_t \times in \times out}$, $b \in \mathbb{R}^{out}$, $c \in \mathbb{R}^{out}$. l is the hidden layer, σ is the activation function and $\otimes$ is the Hadamard product of the two matrices (element-by-element product of the two matrices).

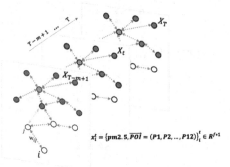

$x_i^t = \{pm2.5, \overline{POI} = (P1, P2, ..., P12)\}_i^t \in R^{f \cdot 1}$

Fig. 3. Graph-structured air quality station data. $X^t = \{x_1^t, x_2^t, \ldots, x_n^t\}$ is the snapshot of all the stations' feature values at time t, including monitoring data and POI data gathered.

4.2 Spatial Dependence Extractor

Extracting inter-station spatial dependence is another key issue for long-term air quality prediction. There is a complex inter-station spatial relationship. Not only the stations' geographical distribution but also the surrounding POI information has great impact. However, these impacting factors (features) are heterogeneous and irregular, since the stations are usually unevenly distributed. Graph convolutional network (GCN) [9], an extension of CNN on non-Euclidean data, is able to extract spatial features of topological graphs and has been successfully used in the fields of natural language processing and image processing. Hence GCN is adopted in this paper as spatial dependence extractor.

The key issue in GCN is the formulation of graph G. To be more specific, we need to assign the weight value for each edge in E to capture the inter-station spatial correlation. We assign an initial value $w_{i,j}$ according to Eq. (3) to edge $e_{i,j}$, representing the relevant degree between station i and j.

$$w_{i,j} = \begin{cases} \exp\left(-\dfrac{\left(e^{ij}\right)^2}{2\alpha}\right), i \neq j \ and \ \exp\left(-\dfrac{\left(e^{ij}\right)^2}{2\alpha}\right) \geq \beta \\ 0, \ \text{otherwise} \end{cases} \quad (3)$$

where e^{ij} is the distance between station i and j, α, β are the thresholds that control the distribution and sparsity of the weigh matrix $W = \{w_{i,j} | i, j \in [1, n]\}$. The initial values of α, β are 5 and 0.5, respectively. Regarding spatial dependence extractor, we choose the first-order Chebyshev polynomial as the graph kernel, and two-layer structured graph convolutional neural network to obtain the spatial dependence. It can be expressed as in Eq. (4).

$$H_t^{(l+1)} = \sigma\left(\widetilde{W} relu\left(\widetilde{W} H_t^{(l)} \theta^l\right) \theta^0\right) \quad (4)$$

where $\tilde{W}$ is the adjacency matrix with self-join added, $\widetilde{W} = \tilde{D}^{-\frac{1}{2}}\tilde{W}\tilde{D}^{-\frac{1}{2}}$ is the normalization operation on the adjacency matrix $\tilde{W}$, $\tilde{D}$ is the degree matrix of $\widetilde{W}$. θ represents the weight matrix of each layer in the spatial dependence extractor, $\theta^l \in R^{h \times h}$, h is the number of neurons in the hidden layer, $\theta^0 \in R^{in \times h}$, in is the model input channels as mentioned in the temporal dependence extractor. The second layer weight matrix $\theta^l \in R^{h \times out}$, out is the output channels. $H_t^{(l)}$ is the feature data for all stations at time t, $t \in \{1, 2, \ldots, m - K_t + 1\}$, generated by temporal dependence extractor and initialized as $H_t^{(l)} = GLU(X)$.

Generally, in our model, we fuse temporal dependence extractor and spatial dependence extractor in one spatiotemporal block, so it can continuously mine spatiotemporal dependence from historical data. By stacking spatiotemporal block, we can obtain long term spatio-temporal dependence and make long short-term prediction.

5 Experimental

5.1 Dataset

We first introduce the dataset used to validate the model performance.

Air Quality Data. The air quality data are measured from 500+ monitoring stations every hour in a city in Eastern China. We used approximately 1,000,000+ records spanning from 2020/10/1 to 2021/1/10.

POI Data. We obtain POI data in the target city by the Amap API [1] with a coverage radius of 500 m and cluster the original 23 POI categories into 12 categories. The POI categories details are shown in Table 1.

5.2 Evaluation Metrics

We use three metrics for performance evaluation, which are defined as:

RMSE (Root Mean Squared Error)

$$RMSE = \sqrt{\frac{1}{\tau n}\sum_{t=T+1}^{T+\tau}\sum_{i=1}^{n}(y_i^t - \hat{y}_i^t)^2}, \tag{5}$$

Table 1. The categories of POI data.

Symbol	Category
P1	Vehicle Services(gas stations, repair)
P2	Transportation spots(Bus stops, car parks, major traffic intersections)
P3	Factories
P4	Decoration and furniture markets
P5	Food and beverage
P6	Shopping malls and Supermarkets
P7	Sports
P8	Parks
P9	Culture and education
P10	Entertainment
P11	Companies
P12	Hotels and estates

MAE (Mean Absolute Error)

$$MAE = \frac{1}{\tau n} \sum_{t=T+1}^{T+\tau} \sum_{i=1}^{n} \left| y_i^t - \hat{y}_i^t \right|, \tag{6}$$

MAPE (Mean Absolute Percentage Error)

$$MAPE = \frac{1}{\tau n} \sum_{t=T+1}^{T+\tau} \sum_{i=1}^{n} \left| \frac{y_i^t - \hat{y}_i^t}{y_i^t} \right|, \tag{7}$$

where y_i^t and $\hat{y}_i^t$ represent the true and predicted values of station i at the future moment $T + t$, respectively. Apparently, smaller RMSE, MAE, MAPE mean better prediction effect.

5.3   Parameter Tuning

The historical window length m is a very important parameter, as it impacts the extent of temporal data we can use. Thus, we conduct an experiment with different m. In this experiment, we vary m as [24 h, 48 h, 72 h, 96 h] and observe the change of prediction error RMSE. As shown in Fig. 4, the prediction error is highest with the lowest m as 24 h and decreases significantly when m increases to 48 h and 72 h. However the RMSE error increases when m further increases to 96 h. This is mainly because when m reaches a certain level, the model complexity increases significantly while the temporal relation becomes increasingly trivial. In summary, $m = 72$ h can yield balanced prediction accuracy and complexity. The experiments on the other two metrics have the similar conclusions. Therefore, we set $m = 72$ h in the experiments hereafter.

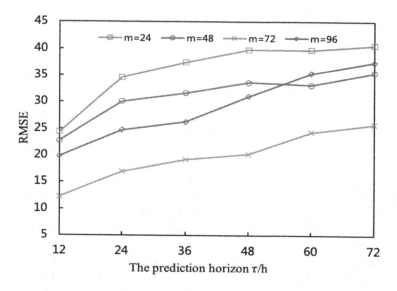

Fig. 4. The model performance under different m

5.4   Results and Discussion

Finally we compare the proposed model with the following baseline methods:

(1) ARIMA [6]: A traditional statistical method in time series prediction.
(2) RNN [12]: A classic deep learning method originally used for natural language processing and time series prediction recently.
(3) LSTM [3]: The most popular variant of RNN, which can capture long-term dependencies via a tricky gating mechanism.
(4) GLU: See Sect. 4.1 for details.
(5) GCN: See Sect. 4.2 for details.
(6) CNN-LSTM [14]: A hybrid model of CNN and LSTM with sequential structure, uses CNN to extract spatial dependence and connects LSTM to extract temporal dependence in air quality data, and finally by using a fully-connected layer to output the prediction results.

Table 2. Performance on short-term prediction

τ	Metric	Method						
		Temporal				Spatial	Hybrid	
		ARIMA	RNN	LSTM	GLU	GCN	CNN-LSTM	Our approach
3 h	RMSE	12.13	8.51	8.68	6.81	9.47	7.39	**6.29**
	MAE	11.62	7.61	7.65	5.33	7.8	5.84	**4.58**
	MAPE (%)	33.62	15.98	15.24	13.89	16.31	15.96	**12.06**
6 h	RMSE	22.9	12.96	12.36	10.29	11.57	10.69	**9.56**
	MAE	18.36	9.54	10.71	7.6	8.38	7.95	**6.47**
	MAPE (%)	38.36	22.28	30.4	18.26	20.3	19.18	**16.39**
9 h	RMSE	28.68	16.32	16.21	12.89	14.37	12.75	**11.92**
	MAE	22.28	11.07	11.75	9.22	10.15	9.08	**7.61**
	MAPE (%)	45.38	27.89	34.71	21.76	24.18	22.41	**17.99**
12 h	RMSE	32.34	18.28	17.95	14.09	16.68	13.02	**12.25**
	MAE	24.79	12.15	13.02	10.66	11.66	9.61	**8.68**
	MAPE (%)	51.83	29.96	28.89	24.7	27.02	24.48	**18.85**

Table 3. Performance on long-term prediction

τ	Metric	Method						
		Temporal				Spatial	Hybrid	
		ARIMA	RNN	LSTM	GLU	GCN	CNN-LSTM	Our approach
24 h	RMSE	35.08	31.76	33.88	24.64	21.2	26.26	**16.95**
	MAE	24.05	24.93	20.88	19.43	24.92	21.42	**11.72**
	MAPE (%)	71.67	38	44.78	36.09	32.68	37.74	**23.47**
36 h	RMSE	32.1	32.11	34.31	30.72	24.7	31.43	**19.19**
	MAE	23.92	22.98	23.09	22.35	17.19	25.46	**13.45**
	MAPE (%)	61.19	34.22	35.1	37.11	36.37	39.56	**29.72**
48 h	RMSE	37.26	33.69	36.95	35.2	26.31	39.84	**20.23**
	MAE	30.54	29.84	30.35	27.3	18.49	27.02	**14.43**
	MAPE (%)	87.94	36.43	62.82	39.03	38.09	35.96	**34.77**
60 h	RMSE	47.65	28.1	43.36	37.69	28.1	45.46	**22.38**
	MAE	23.6	23.23	36.64	27.46	19.87	30.53	**16.72**
	MAPE (%)	50.14	50.31	45.86	42.23	44.19	43.21	**42.05**
72 h	RMSE	63.02	49.67	52.47	46.97	47.67	49.93	**25.81**
	MAE	53.91	36.73	43.65	30.23	31.83	32.74	**18.57**
	MAPE (%)	100.52	53.72	105.05	47.45	46.98	48.35	**40.56**

As shown in Table 2 and Table 3, the proposed model achieves the lowest errors for both short and long-term predictions. From the tables we can find that as the prediction time horizon τ increases, all metrics RMSE, MAE and MAPE gradually increase. The following are detailed discussions regarding high prediction precision and long-term prediction.

High Prediction Precision. We can find that prediction models based on hybrid deep learning models, including our proposed model and CNN-LSTM, which emphasises spatiotemporal dependence, typically have better prediction accuracy than the others that only consider one dimensional dependence (temporal or spatial dependence) (e.g. ARIMA, RNN, LSTM, GLU and GCN). For example, for 12 h prediction, compared to ARIMA, RMSE of our proposed model is reduced by approximately 62.12%. This is because ARIMA is suitable for the data with strong autocorrelation while air quality data are influenced by multivariate factors and hence have strong randomness. Compared with LSTM, the RMSE errors of our model and GLU are reduced by 31.75% and 21.5% respectively. This is mainly because the chain structure of RNN and its variant are not suitable for capturing long term dependence.

Compared with GCN, the RMSE error of our model is reduced by 5.9%. The lower prediction performance of GCN is mainly because GCN only captures the spatial dependence while ignoring that the air quality is also time series data.

As to the CNN-LSTM model with spatiotemporal dependence extraction, our model reduces RMSE error by up to 26.56%. One possible reason is CNN does not work well with non-Euclidean spatial data.

For the experiments with the other τ values, we can obtain similar conclusions. Overall, the experimental results show that with varying prediction time horizon τ, our approach outperforms the other models, mainly thanks to the fusion structure of our model to extract spatiotemporal features coherently.

Performance on Long-Term Dependence. When we focus on the long-term prediction, i.e. $\tau = 72$ h, we can find that the advantage of the proposed model is more obvious. For example, RMSE is only 54.95% that of the other models. This demonstrates that our model can truely mine the spatiotemporal correlation for long-term prediction.

6   Conclusion

Considering that the air quality data is characterised by dynamic variability and spatiotemporal correlation, we exploit urban heterogeneous big data and propose a spatiotemporal merged model for long short-term air quality prediction. The proposed model is a complete convolutional structure and consists of stacked spatiotemporal fusion blocks, each integrates a GLU for extracting long-term temporal dependence and a GCN for extracting spatial dependence. Using such a fusion structure, the spatiotemporal correlations of air quality data are extracted coherently, reducing the transmission errors generated by existing models through separate models. We have conducted a bunch of experiments, and compared with 6 baseline models such as LSTM model, CNN-LSTM and GCN. The results are consistent with our expectations, our model outperforms the other models in terms of both long short-term predictions. In summary, our model successfully captures the long-term spatiotemporal dependence of urban air quality data, and it can also be applied to other long-term prediction tasks for spatiotemporal data.

References

1. https://lbs.amap.com/api/webservice/guide/api/search/
2. Bai, Y., Li, Y., Wang, X., Xie, J., Li, C.: Air pollutants concentrations forecasting using back propagation neural network based on wavelet decomposition with meteorological conditions. Atmos. Pollut. Res. **7**(3), 557–566 (2016)
3. Chang, Y.S., Chiao, H.T., Abimannan, S., Huang, Y.P., Tsai, Y.T., Lin, K.M.: An LSTM-based aggregated model for air pollution forecasting. Atmos. Pollut. Res. **11**(8), 1451–1463 (2020)
4. Cheng, W., Shen, Y., Zhu, Y., Huang, L.: A neural attention model for urban air quality inference: learning the weights of monitoring stations. In: Proceedings of the AAAI Conference on Artificial Intelligence, vol. 32 (2018)
5. Dauphin, Y.N., Fan, A., Auli, M., Grangier, D.: Language modeling with gated convolutional networks. In: International Conference on Machine Learning, pp. 933–941. PMLR (2017)
6. Díaz-Robles, L.A., et al.: A hybrid ARIMA and artificial neural networks model to forecast particulate matter in urban areas: the case of Temuco, Chile. Atmos. Environ. **42**(35), 8331–8340 (2008)
7. Fu, R., Zhang, Z., Li, L.: Using LSTM and GRU neural network methods for traffic flow prediction. In: 2016 31st Youth Academic Annual Conference of Chinese Association of Automation (YAC), pp. 324–328. IEEE (2016)
8. Huang, W., Li, T., Liu, J., Xie, P., Du, S., Teng, F.: An overview of air quality analysis by big data techniques: monitoring, forecasting, and traceability. Inf. Fusion **75**, 28–40 (2021)
9. Kipf, T.N., Welling, M.: Semi-supervised classification with graph convolutional networks. arXiv preprint arXiv:1609.02907 (2016)
10. Liang, Y., Ke, S., Zhang, J., Yi, X., Zheng, Y.: GeoMAN: multi-level attention networks for geo-sensory time series prediction. In: IJCAI, vol. 2018, pp. 3428–3434 (2018)
11. Liu, B.C., Binaykia, A., Chang, P.C., Tiwari, M.K., Tsao, C.C.: Urban air quality forecasting based on multi-dimensional collaborative Support Vector Regression (SVR): a case study of Beijing-Tianjin-Shijiazhuang. PLoS ONE **12**(7), e0179763 (2017)
12. Liu, B., Yan, S., Li, J., Li, Y., Lang, J., Qu, G.: A spatiotemporal recurrent neural network for prediction of atmospheric PM2.5: a case study of Beijing. IEEE Trans. Comput. Soc. Syst. **8**(3), 578–588 (2021)
13. Lv, Z., Xu, J., Zheng, K., Yin, H., Zhao, P., Zhou, X.: LC-RNN: a deep learning model for traffic speed prediction. In: IJCAI, vol. 2018, p. 27 (2018)
14. Qin, D., Yu, J., Zou, G., Yong, R., Zhao, Q., Zhang, B.: A novel combined prediction scheme based on CNN and LSTM for urban $PM_{2.5}$ concentration. IEEE Access **7**, 20050–20059 (2019)
15. Wang, S., Li, Y., Zhang, J., Meng, Q., Meng, L., Gao, F.: PM2.5-GNN: a domain knowledge enhanced graph neural network for PM2.5 forecasting. In: Proceedings of the 28th International Conference on Advances in Geographic Information Systems, pp. 163–166 (2020)
16. Xu, J., Chen, L., Lv, M., Zhan, C., Chen, S., Chang, J.: HighAir: a hierarchical graph neural network-based air quality forecasting method. arXiv preprint arXiv:2101.04264 (2021)

17. Yi, X., Zhang, J., Wang, Z., Li, T., Zheng, Y.: Deep distributed fusion network for air quality prediction. In: Proceedings of the 24th ACM SIGKDD International Conference on Knowledge Discovery & Data Mining, pp. 965–973 (2018)
18. Zheng, Y., Liu, F., Hsieh, H.P.: U-Air: when urban air quality inference meets big data, pp. 1436–1444 (2013)

Colour Coherence Entropy for Ceramic Fragment Analysis

Buwei He[1,2], Yuhang Gao[3], and Yi Sun[3(✉)]

[1] School of Computing Science, University of Glasgow, Glasgow G12 8QQ, UK
[2] International School, Beijing University of Posts and Telecommunications, Beijing 100876, China
[3] School of Computer Science (National Pilot Software Engineering School), Beijing University of Posts and Telecommunications, Beijing 100876, China
sunyisse@bupt.edu.cn

Abstract. Thousands of broken porcelain tiles are unearthed every year and it is difficult to sort out the individual pieces from the jumble of tiles. Digital restoration of ancient artifacts reduces the reliance on heavy human labour for identification in the restoration process. Image recognition technology is one of the effective means of digital restoration, and the ability to extract the features of the image determines the recognition result. In this work, we aim to cluster a given collection of pieces into sets that came from different caremic objects. Features are extracted from each piece and compared to perform clustering. Color coherence vectors (CCV) put the consistency of the same color into consideration, add spatial information to improve the performance of image recognition, which provide an idea of partition for content-based image retrieval (CBIR). This paper proposed an runtime improved CCV, colour coherence entropy (CCE), based on CCV and entropy. CCV divides the quantized colors into coherent part and incoherent part. The entropy is used to describe the distribution of those two parts to reduce the dimensionality of the data. As a hybrid feature of image retrieval, CCE reduces the data dimension from the original 64 or even higher to 2, while preserving the original features as much as possible through the entropy function. In a database of 10,000 images, CCE was 94.3% faster than CCV in retrieval computation efficiency. In 93 clustered image datasets, the retrieval accuracy of CCE was 25.1% higher than that of the one-dimensional entropy difference method. Considering the balance between computational efficiency and retrieval accuracy, CCE has a good performance for image retrieval pre-processing in CBIR and is suitable for massive image retrieval for ceramic fragment analysis.

Keywords: Content-based Image Retrieval · Colour Entropy · Colour Coherence Vectors · Colour Coherence Entropy

S. Yang and S. Islam (Eds.): APWeb-WAIM 2022 Workshops, CCIS 1784, pp. 246–259, 2023.
https://doi.org/10.1007/978-981-99-1354-1_21

1 Introduction

Fragments restoration of antique ceramics for both aims of ornamental and archaeological value has a high demand in China [1, 2]. The whole procedure of restoration involves several levels and steps. The identification of antique ceramics fragments is one of the critical steps of restoration that requires significant investment [1, 3]. This work tried to focus on the initial screening of identification. The digital restoration of cultural objects already has some workable technical routes. For example, the 3D scanning point cloud technique has yielded promising results in the restoration of terracotta [4] and bronze [5]. 3D laser scanning has been used to repair broken ceramic pottery and even fill in the missing parts [6]. Since precise data on the contour characteristics of the fragments can be obtained by the 3D scanning method, it is suitable for the restoration of individual pieces but not for the initial screening of large numbers of fragments. Therefore, the stereo profile of the fragments can be temporarily disregarded at the initial stage of identification.

Using features from 2D images of fragments for comparison is a more efficient method. Content-based image retrieval (CBIR) method is well suited to 2D image retrieval from large data sets [7, 8]. CBIR enables the comparison of visual similarities between images by extracting image features [8]. Generally, the features extracted can be divided into 1) global features that represent the whole image, and 2) local features that focus on some parts of images. Global features allow for faster feature extraction and similarity calculation. Local features are more suitable for retrieval in complex scenes or object recognition [9].

As a refinement of the colour histogram, colour coherence vectors (CCV) [10] is a kind of colour-based image retrieval method of CBIR. CCV introduces spatial information by splitting histograms according to the coherence of pixels. Data dimensions in calculation and retrieval phases of CCV are determined by the colour quantisation level of an image, which negatively affects its performance [11].

The introduce of colour entropy provided a measurement of the colours dispersion in a image, which can be used as a dimensionality reduction method to improve the performance of CCV. This is also the motivation for this work.

This paper focuses on a hybrid colour-based image retrieval algorithm for image comparing. Entropy was introduced to calculate the coherent and incoherent parts of a histogram splited by the CCV method, reducing the original data dimension to two, called colour coherence entropy (CCE). CCE method was compared with CCV, Improved CCV (ICCV) [12], and entropy difference (ED) method together in a database of 10,000 images for the computational efficiency and in a database of 93 images for the retrieval correctness rate. The former database came from the test batch of CIFAR-10[1] [13], and the latter database came from shooting clips, websites and Yu Yong's book [14]. The latter database is shown in Appendix A.

---

[1] The CIFAR-10 data set consists of 60000 32×32 colour images that can be sorted into several classes, such as airplane, automobile, bird, cat, etc. CIFAR-10 has five training batches and one test batch. The test batch consists of ten thousand images.

2 Background

2.1 Colour Histogram

Values are grouped into bins by colour histogram. As a global feature of the image, the histogram is translation, scale and rotation invariant [15]. The similarity between two histograms can be compared with histogram intersection distance [16] or Chi-square Statistic [17], etc. Lack of spatial information is a disadvantage of histograms, images with completely different content may have the same histogram, which can lead to retrieval failure [9]. So the recognition performance can be improved by adding spatial information [10].

2.2 Colour Coherence Vectors

The spatial information of pixels is added in CCV. CCV splits the histogram based on spatial coherence [10]. The idea of CCV is to divide the pixels of each bin of the histogram into two parts: if some pixels in the bin occupy a contiguous area larger than a threshold, then the pixels are coherent. Otherwise, the pixels are incoherent. By extracting successive large blocks of colour, haphazardly distributed pixels are separated from an image. The colour coherence vector can be presented as Eq. (1).

$$CCV = \langle (\alpha_1, \beta_1), \ldots, (\alpha_i, \beta_i), \ldots, (\alpha_n, \beta_n) \rangle \tag{1}$$

Where α_i is the number of coherent pixels of quantised colour i, β_i is the number of incoherent pixels.

The introduction of spatial information improves the retrieval accuracy of CCV [10], but the additional histogram splitting operation makes retrieval efficiency poor. An improved CCV method (ICCV) was proposed by recording the coordinate of the central pixel of every max-coherent region colour [12]. The introduction of coordinate made ICCV sensitive to the rotation of the image compared to CCV and the CCE proposed in this paper. This feature is detrimental in specific usage scenarios, such as the rotation angle of the image to be retrieved is not known in advance.

2.3 Entropy of Images

Entropy is a measure of the information volume of an image [18]. A digital image contains information, the amount of which can be measured by the probability of different colours appearing on pixels. An increase in the entropy of an image corresponds to an increase in the level of uncertainty of the image pixels [19]. Images with the same information entropy can be considered to convey the same amount of information and contain similar content. Since histograms can be converted into probability distributions [18, 20], and each bin of the histogram relates to a probability distribution value. Entropy can be obtained by Eq. (2).

$$H_I = -\sum_{i=1}^{n} P_i lb P_i, P_i = \frac{h_i}{N}, N = \sum_{i=1}^{n} h_i \tag{2}$$

where P_i indicates the probability that colour i appears in image.

The entropy difference (ED) was used to compare the similarity of the image information, which was used in the initial screening stage of the entropy enhanced L1 norm (EELN) algorithm to narrow the result set and improve the retrieval efficiency [18]. Entropy can be used as a data dimension reduction method to decrease the computation time [21].

3 Colour Coherence Entropy

3.1 HSV Colour Space

This work began with the choice of colour space. The HSV colour space is a non-linear system for describing the colour in terms of hue, saturation and luminance [22]. The hue component is extracted directly in HSV, allowing for a uniform perception of colour differences and reducing the effects of illumination [16, 23]. This feature reduces the sensitivity of features to brightness and reduces the interference of scene lighting in recognition of picture content [24].

3.2 Colour Non-uniformly Quantification

Hue is the focus of colour-based retrieval methods and should be given a higher quantisation level than the saturation and luminance components and has a higher weight when synthesizing feature vectors [25]. In addition, most of the primary hue pixels are concentrated in certain intervals, the size of the quantization interval varies according to the degree of concentration of the hue. [26]. In this work, 16 slots were gotten for the hue attribute and 4 slots for the saturation attribute and luminance attribute respectively.

3.3 Compute CCE

The feature dimension and feature computation time were reduced by converting the histogram into a probability distribution to get the information entropy in CCE. Furthermore, spatial information was introduced by calculating the information entropy of the two sets of CCV (coherent and incoherent vectors) separately. Thus the entropy of central and scatter information in the image was stored in CCE. The CCE of an image can be obtained as follows.

Procedure

Input: A greyscale image through Gaussian low-pass filter.

Output: The CCE of image.

Begin

   # Get the colour histogram

   initiate every bin in histogram

   for every pixel p in image

      if p <= bin[j].threshold

        bin[j] += 1

   # Divide every bin into two parts according to the coherence

   for every bin in histogram

      the number of pixels in bin[j] that are adjacent is m

      if m > bin[j] * 0.01

        bin_coherent[j] += m

      else

        bin_incoherent[j] += m

   # Convert the histogram into a probability distribution

   $\text{p\_coherent} = \langle p_1, \ldots, p_n \rangle, \; p_i = \frac{\text{bin\_coherent[j]}}{\sum \text{bin\_coherent}}$

   $\text{p\_incoherent} = \langle p_1, \ldots, p_n \rangle, \; p_i = \frac{\text{bin\_incoherent[j]}}{\sum \text{bin\_incoherent}}$

   # Calculate entropy

   $\text{H\_coherent} = \sum_{i=1}^{n} -p_i \, log_2(p_i), \; p_i \in p\_coherent$

   $\text{H\_incoherent} = \sum_{i=1}^{n} p_i \, log_2(p_i), \; p_i \in p\_incoherent$

   return $\langle \text{H\_coherent}, \text{H\_incoherent} \rangle$

End Procedure.

3.4 Compare CCE

When two images I and I' picked up to compare by CCE, calculate their coherent entropy and incoherent entropy. $E = (H_\alpha, H_\beta), E' = \left(H'_\alpha, H'_\beta \right)$. The similarity between the two images can compute as $\Delta_E = \left| H_\alpha - H'_\alpha \right| + \left| H_\beta - H'_\beta \right|$.

4 Results and Discussion

The programme was operated on personal computer with environment as followed. Host software: MATLAB R2021b. Processor: Four Intel(R) Core(TM) i7-7500U CPU @ 2.70 GHz. RAM: 16.0 GB. Display adapter: Intel(R) HD Graphics 620. OS: Windows 10.

The performance of the CCE algorithm was evaluated in two ways. 1) Runtime performance: how long it takes to compute feature values for a batch of images & how long it takes to retrieve the highest similarity image with the query image. 2) Retrieval performance: given a set of labelled images, it has better clustering results. The specific methodology for evaluating retrieval performance is described in detail in Retrieval Performance section.

The criteria in runtime performance was proposed on data sets derived from CIFAR-10. The data set for the last criterion consists of images taken from the shooting clips, website and book illustrations. In all experiments, HSV colour space was used, and the number of quantisation steps was 256. A rotationally symmetric 5×5 Gaussian low-pass filter with a standard deviation of 2 was used for denoising before colourful images were converted to grey images. For algorithms involving coherent division characteristics (CCV, ICCV & CCE), the coherence threshold was uniformly set to 1%, i.e. if the number of pixels in a connected component was greater than or equal to 1% of the number of pixels in the entire image, the region was considered coherent.

4.1 Runtime Performance

The average running time taken by each method to compute features in different databases is shown in Fig. 1 (left). The specific data of Fig. 1 (left) was shown in Table 1. The ten thousand images from the test batch in CIFAR-10 are used as a test set. Images from the test set were randomly selected to make up seven different sizes of database (consisting of 100, 213, 467, 1000, 2137, 4677 and 10000 pieces of the image, respectively). Though every image has been labelled according to the classes, they were treated as natural image set without concern about the actual classes here. Note that the coordinate of the figure is exponentially distributed.

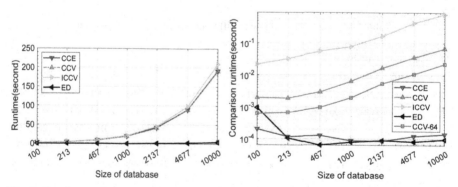

Fig. 1. Runtime performance (left) & Comparison runtime (right) performance of four methods in different sizes of data set.

All four methods (CCE, CCV, ICCV, ED) are histogram-based. So the four methods take identical time to calculate the histogram. The entropy was calculated directly from the histogram by ED, which got the best performance. Among the other three methods that require splitting the histogram based on coherence, CCV and CCE performed almost

Table 1. Runtime in different size of databases (second).

Methods	Size of databases						
	100	213	467	1000	2137	4677	10000
CCE	2.031045	4.169263	9.116133	19.586228	41.980043	89.630411	191.721367
CCV	1.933340	4.051007	8.992701	19.044840	40.984114	89.411898	191.020966
ICCV	2.062009	4.522616	9.845123	20.914140	44.909465	97.961443	209.771679
ED	0.083084	0.239385	0.2841964	0.0431446	1.0010184	2.1278964	4.4411455

identically. The coordinates of the centre of the coherent region had to be calculated in ICCV, which got the worst performance.

Another runtime performance was the comparative efficiency of the feature values. The way to construct databases was the same as the last one. Ten query images were randomly selected in each database. The average comparison time was recorded for the most similar image in each database. Typically, a relatively small number of quantisation levels often be chosen in CCV [10]. Accordingly, the performance of CCV with 64 quantisation levels (labelled by *CCV-64*) was tested in order to match the practical application scenario. The result is depicted in Fig. 1 (right) and the specific data of Fig. 1 (right) was shown in Table 2.

CCE and ED had low dimensionality, and their comparison runtimes were almost identical within an order of magnitude of each other (all lower than 0.137 ms for 10,000 images). The speed of CCE was around two orders of magnitude faster than CCV (67.7 ms for 10,000 images). CCV-64 (22.3 ms for 10,000 images) was slightly faster than CCV. ICCV had the worst performance (809.6 ms for 10,000 images) that was far away from the others.

Table 2. Comparison runtime in different size of databases (second).

Methods	Size of databases						
	100	213	467	1000	2137	4677	10000
CCE	0.00097544	0.00006795	0.00007213	0.00009962	0.00014068	0.00011464	0.00013674
CCV	0.00194497	0.00187981	0.00294608	0.00637336	0.01650806	0.03519341	0.06774324
CCV-64	0.00062548	0.00066966	0.00097173	0.00195317	0.00542872	0.01072382	0.02234505
ICCV	0.02152891	0.03188618	0.05559294	0.07703617	0.16575843	0.42568559	0.80962933
ED	0.0009533	0.0001048	0.00006413	0.00007912	0.00008998	0.00008215	0.00009685

Note that as images in the test set were selected randomly, the two images with the highest similarity of output may not have actual content consistency, therefore, the correctness of the retrieval results was not considered in this section, but in the next sectionNote that as images in the test set were selected randomly, the two images with the highest similarity of output may not have actual content consistency, therefore, the correctness of the retrieval results was not considered in this section, but in the next section.

4.2 Retrieval Performance

The retrieval test data set was sourced from a wide range of sources and contained 93 images, which were 64×64 colour JPG images taken from ceramic glazes. Part of the images is shown in Fig. 2. The number before the underscore represents the grouping of the image, and the number after the underscore represents the number of the image within the category.

The colour and distribution of the patterns in the ceramic glaze had a particular pattern, and the images were manually labelled based on whether coming from the same utensil. After the algorithm extracted the CCE features of each image, clustering algorithms were used to test the clustering performance. Note the algorithm such as K-means may not be able to cover all the correct categories when selecting new clusters randomly; and there is a risk of falling into a local optimum solution. To exclude the effect of clustering algorithm stability on feature performance, the following evaluation method was conducted.

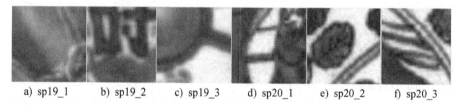

a) sp19_1   b) sp19_2   c) sp19_3   d) sp20_1   e) sp20_2   f) sp20_3

Fig. 2. Part of the image set for clustering evaluation [14].

In a data set, each entry belongs to a unique category, each category having no less than two data. A query image is selected, and a collection (retrieval domain) is generated according to the feature extraction algorithm. The number of entries belonging to other classes in the collection needs to be as small as possible while containing all the target entries of the same class as the query image.

Step 1: A query image I is randomly selected from data set $D = \{I_1, I_2, ..., I_n\}$. The rest of the images can be divided into $C_I = \{I_i| I_i$ in the same category as I$\}$ and $C_{I'} = \{I_i| I_i$ in different category as I$\}$.

Step 2: Calculate the distance d = max dist(I, C_I) from the farthest target image to I. The area within d is the retrieval domain $\mathcal{D}_I$ of I. All images contained in $\mathcal{D}_I$ form a set $C_{\mathcal{D}_I}$, $C_I \subseteq C_{\mathcal{D}_I}$. Note that $\mathcal{D}_I$ may not be the minimum retrieval domain for the category of I, and this effect will be reduced by increasing the size of the data set and random sampling.

Step 3: percision $= \frac{|C_I|}{|C_{\mathcal{D}_I}|}$. The accuracy rate of the algorithm can be calculated by this method that simplifies the evaluation process with the artificial elimination of false-negative sets.

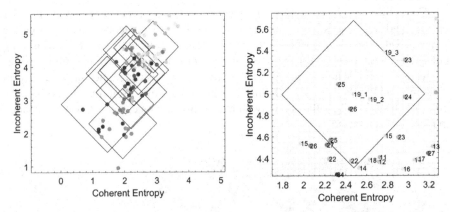

Fig. 3. Left: Twenty randomly selected points and their query domains. Right: Retrieval domain for an image of category No. 19

The above steps were performed twenty times on CCE with the retrieval test data set to obtain the result shown in Fig. 3 (left). Each point in the graph represents an image in the data set. The distance between points reflects the similarity of the two images. The colour of each point was generated based on their real category. The squares in the diagram represent the retrieval domain for each query image.

The image "sp19_1.jpg" (can be found in Fig. 2) will be used as an example to illustrate the process of calculating the precision of the algorithm. Figure 3 (right) is an enlarged partial display of Fig. 3 (left). The square shown in the figure is the retrieval domain belonging to "19_1". There were a total of three images in the data set that fall under category No. 19; "19_3" is the furthest target image from "19_1". In order to encompass "19_3", the retrieval domain had to contain five images from other categories. Thus, the precision of this query was *percision* $= \frac{3}{8} = 0.375$. After twenty random queries, the retrieval performance of ED and CCE is shown in Fig. 4. The average retrieval accuracy of CCE in the clustered image data set was 25.1% higher than that of ED. Essentially, by mapping a one-dimensional point set to a two-dimensional space somehow, differences between collections can be better quantified.

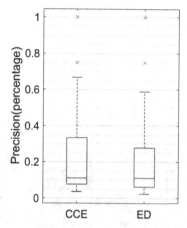

Fig. 4. The retrieval performance of ED and CCE.

5 Conclusion

This work proposed CCE, a hybrid feature extraction method based on CCV and entropy, to reduce the computational cost when comparing features in the initial stage of fragment identification. Although the computation effort was increased when calculating the features compared with ED, it is worth retaining more spatial information. Mapping a low-dimensional vector to a higher dimension is intuitively counter-optimal to performance unless the operation results in worthwhile new data features. Experimental results shown that our method was comparable to ED in speed, outperformed other methods in retrieval performance. It is hoped to balance computational speed and accuracy. At the same time, the development of a database of ceramic fragment image is imminent to meet the testing requirements of the fragment recovery algorithms.

Appendix A Cluster Image Set

There are 93 64 × 64 colour JPG images. Twenty-two of these are close-up photographs from ceramic pot I bought, eight from copyright-free images on the internet and rest from book images.

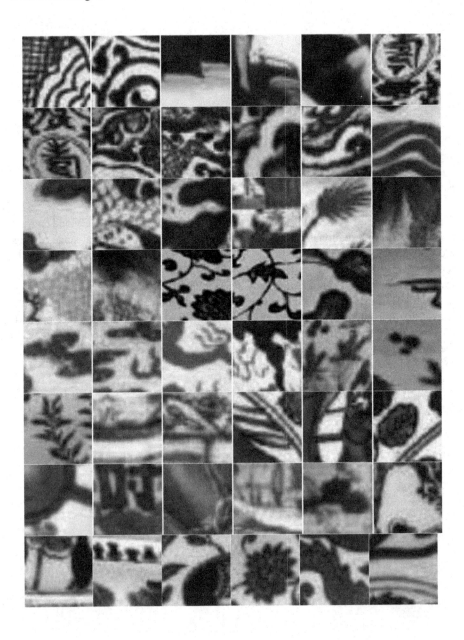

References

1. Sisu, R.: Shock! Broken porcelain is more precious than intact. Imperial Kiln in Ming Dynasty. http://www.360doc.com/content/15/0920/12/6140124_500268487.shtml. Accessed 20 May 2022
2. Xinhuanet: Jingdezhen imperial kiln ancient porcelain fragments "restoration" ceramic legend. http://collection.sina.com.cn/cpsc/2018-10-19/doc-ifxeuwws5904114.shtml. Accessed 20 May 2022

3. Bjnews: How to spell 100000 pieces of porcelain? Come to the old summer palace to see and repair cultural relics. https://baijiahao.baidu.com/s?id=1645188641728414185&wfr=spi der&for=pc. Accessed 20 May 2022

4. Fu-Qun, Z., Ming-Quan, Z.: Effective blocks matching algorithm of terracotta warrior, vol. 26, no. 03, pp. 198–203 (2017)

5. Mingqiang, W., Honghua, C., Yangxing, S., Jun, W., Yanwen, G., Xuefeng, Y.: Digital restoration of damaged cultural relics: a case study on Chinese unearthed bronzes, vol. 33, no. 05, pp. 789–797 (2021)

6. Fragkos, S., Tzimtzimis, E., Tzetzis, D., Dodun, O., Kyratsis, P.: 3D laser scanning and digital restoration of an archaeological find. In: 2018 22nd International Conference on Innovative Manufacturing Engineering and Energy (IManE&E), Chisinau, Moldova, vol. 178 (2018)

7. Raghunathan, B., Acton, S.T.: A content based retrieval engine for circuit board inspection. In: Proceedings 1999 International Conference on Image Processing (Cat. 99CH36348), vol. 1, pp. 104–108 (1999)

8. Ali, F., Hashem, A.: Content based image retrieval (CBIR) by statistical methods. Baghdad Sci. J. **17**(2), 694–700 (2020)

9. Hameed, I.M., Abdulhussain, S.H., Mahmmod, B.M.: Content-based image retrieval: a review of recent trends. Cogent Eng. **8** (2021)

10. Pass, G., Zabih, R.: Histogram refinement for content-based image retrieval. In: 3rd IEEE Workshop on Applications of Computer Vision (WACV 1996), Sarasota, Fl, pp. 96–102 (1996)

11. Weber, R., Schek, H.-J., Blott, S.: A quantitative analysis and performance study for similarity-search methods in high-dimensional spaces. In: VLDB, vol. 98, pp. 194-205 (1998)

12. Chen, X., Gu, X., Xu, H.: An improved color coherence vector method for CBIR, pp. 33–37 (2008)

13. Krizhevsky, A.: Learning multiple layers of features from tiny images (2009)

14. Yong, Y.: Ceramic underglaze celadon ware decoration. Jiangsu Phoenix Art Publishing House Publication (2018)

15. Shrivastava, N., Tyagi, V.: An efficient technique for retrieval of color images in large databases. Comput. Electr. Eng. **46**, 314–327 (2015)

16. Swain, M.J., Ballard, D.H.: Color indexing. Int. J. Comput. Vis. **7**(1), 11–32 (1991). https://doi.org/10.1007/BF00130487

17. Cha, S.-H.: Comprehensive survey on distance/similarity measures between probability density functions. Int. J. Math. Model. Meth. Appl. Sci. **1** (2007)

18. Zachary, J., Iyengar, S.S.: Information theoretic similarity measures for content based image retrieval. J. Am. Soc. Inf. Sci. Technol. **52**(10), 856–867 (2001)

19. Susanj, D., Tuhtan, V., Lenac, L., Gulan, G., Kozar, I., Jericevic, Z.: Using entropy information measures for edge detection in digital images. In: 2015 38th International Convention on Information and Communication Technology, Electronics and Microelectronics (MIPRO), Opatija, Croatia, pp. 352–355 (2015)

20. Martinez-Aroza, J., Gomez-Lopera, J.F., Blanco-Navarro, D., Rodriguez-Camacho, J.: Clustered entropy for edge detection. Math. Comput. Simul. **182**, 620–645 (2021)

21. El-Sayed, M.A.: A new algorithm based entropic threshold for edge detection in images. Int. J. Comput. Sci. Issues **8**(5) (2011). Article no. 1

22. Vera, E., Torres, S.: Adaptive color space transform using independent component analysis. In: Conference on Image Processing - Algorithms and Systems V, San Jose, CA, vol. 6497 (2007)

23. Dong, Y.X.: Image retrieval based on HSV feature and edge direction feature. In: 2015 International Conference on Advances in Mechanical Engineering and Industrial Informatics (AMEII), Zhengzhou, People's Republic of China, vol. 15, pp. 1002–1007 (2015)

24. Danapur, N., Dizaj, S.A.A., Rostami, V.: An efficient image retrieval based on an integration of HSV, RLBP, and CENTRIST features using ensemble classifier learning. Multimed. Tools Appl. **79**(33–34), 24463–24486 (2020). https://doi.org/10.1007/s11042-020-09109-9
25. Liu, J., Zhao, H., Kong, D., Chen, C.: Image retrieval based on weighted blocks and color feature. In: 2011 International Conference on Mechatronic Science, Electric Engineering and Computer (MEC), pp. 921–924 (2011)
26. Sajjad, M., Ullah, A., Ahmad, J., Abbas, N., Rho, S., Baik, S.W.: Integrating salient colors with rotational invariant texture features for image representation in retrieval systems. Multimed. Tools Appl. **77**(4), 4769–4789 (2018). https://doi.org/10.1007/s11042-017-5010-5

A Fast Method of Legal Decision Recommendation System

Jing Zhou, Yong Li, Bin Zhang, Xin Chen, Xiaopeng Wei, and Ximin Sun[✉]

State Grid Ecommerce Technology Co., Ltd., Tianjin, China
sunsemon@126.com

Abstract. Current engines search based on keywords, but in the process of legal case search recommendation, only some keywords are not enough. This work mainly completes the classification task of legal texts in three aspects: crime prediction, articles of law recommendation, and prison term prediction. We use TextCNN as a baseline. First of all, we make different attempts on the receptive field of the convolutional layer to verify its impact on the task of this project. Secondly, in order to better reflect the impact of context on the task of this project, we combined TextCNN and bidirectional GRU or Attention. In order to verify the effect of the above method, we use the CAIL2018 dataset for verification. The dataset contains more than 2.6 million cases. The experimental results show that the change of the receptive field and the introduction of attention have a specific impact on this project, and the combination of TextCNN and Attention has also achieved good results for the comprehensive performance of the three tasks of this work.

Keywords: Legal Decision Recommendation · Attention Mechanism · Text Classification

1 Introduction

Deep learning technology extracts and filters the features of input information and data layer by layer by establishing artificial neural network, and finally obtains the results of tasks such as classification and prediction, which has completely changed many machine learning tasks. The success of Deep Learning (DL) [1] has spread to a variety of domains such as computer vision, natural language processing, audio analysis and so on.

Text Classification is a category of Natural Language Processing (NLP) tasks with real-world applications such as spam, fraud, and bot detection [2–4], emergency response [5], and commercial document classification, such as for legal discovery [6]. Deep learning based models have surpassed classical machine learning based approaches in various tasks of Text Classification (TC), including sentiment analysis, news categorization, question answering and so on. Sentiment analysis is the task of analyzing people's opinions in textual data, and extracting their polarity and viewpoint. News Categorization is the task of identifying

S. Yang and S. Islam (Eds.): APWeb-WAIM 2022 Workshops, CCIS 1784, pp. 260–268, 2023.
https://doi.org/10.1007/978-981-99-1354-1_22

emerging news topics or recommending relevant news based on user interests question answering. Given a question and a set of candidate answers (e.g., text spans in a document in SQuAD [7]). In this work, we focus on the classification tasks of legal texts in three aspects: crime prediction, articles of law recommendation, and prison term prediction.

We pay attention to this field because of the following situation: with the development of society and the increase in economic activities, the number of legal disputes and incidents is increasing. But the number and cost of professional lawyers are unbearable for ordinary people. Most people use search engines for information queries, but the quality of search engine answers varies, even wrong. At the same time, artificial intelligence is also booming. In particular, deep neural networks have made remarkable achievements. This gave birth to a combination of neural networks and legal text processing. In recent years, a large number of legal case datasets have been made public, which has also become the basis for integration.

Our work is inspired by Legal Judgment Prediction (LJP), which is a traditional task in the combination of artificial intelligence and laws. It aims to train a machine judge to predict the judgment results automatically according to the facts. A well-performed LJP system can not only benefit those who are not familiar with laws but also provide a reference to professionals, e.g., lawyers and judges. We exploit a large-scale criminal dataset constructed from Chinese law documents called CAIL2018 [8]. CAIL2018 contains more than 2.6 million criminal cases published by the Supreme People's Court of China, which are several times larger than other datasets in existing works on judgment prediction. Based on this dataset, we attempt to explore the more effective methods to deal with the proposed three tasks.

The contributions of this paper are summarized as follows:

- we make different attempts on the receptive field of the convolutional layer to verify its impact on the task of this project.
- We introduce attention mechanism into the baseline model and compare it with bidirectional GRU. We verify the performance of these two model structures for different tasks. Our proposed method has better performance in parameters and Flops and also achieved higher accuracy.
- We have verified our method on CAIL2018 dataset. The experimental results confirm that our method achieves higher accuracy in the classification task of legal texts than the previous methods.

2   Related Work

2.1   TextCNN Model

When it comes to CNN, it is usually considered to belong to CV and is used for computer vision work. However, Yoon Kim made some changes to the input layer of CNN and proposed the text classification model TextCNN [9]. TextCNN model is a variant of the CNN model. In addition to retaining the characteristics of the original CNN, it also adds the ability to extract text features. TextCNN uses one-dimensional convolution to obtain the n-gram feature representation of the sentence, By adjusting the height of the convolution kernel, TextCNN can flexibly process various timing information of the comprehensive vocabulary, which improves the model's ability to interpret the text.

2.2 Graph Convolutional Networks for Text Classification

In recent years, GNNs have attracted wide attention [10,11]. Kipf and Welling [12] presented a simplified graph neural network model, called graph convolutional networks (GCN), which achieved state-of-the-art classification results on a number of benchmark graph datasets. GCNs have recently been used in text classification since GNNs can model complex semantic structures and perform well in handling complex network structures [13,14]. TextRank [15] was the earliest graph-based model that applies graph structures to NLP, representing a natural language text as a graph. Then, some methods transform the text classification problem into the document node classification problem in the large graph, such as TextGCN [16], HyperGAT [17], and TensorGCN [18]. The other method generates small individual graphs for each document in the corpus, such as semantic and syntactic dependency graphs. The words of each document are the nodes of the graph and convert the text classification problem using a graph classification problem, such as S-LSTM [14] and TextING [19]. However, Some contextual information will be lost in the process of converting text data into graph structure data, and the propagation of information between nodes will introduce a large number of learnable parameters. These defects greatly reduce the reasoning speed of the model.

2.3 Attention Mechanism

Attention mechanism [20] has been used as an important component across a wide range of NLP models. Typically, an attention layer produces a distribution over input representations to be attended to. Such a distribution is then used for constructing a weighted combination of the inputs, which will then be employed by certain downstream modules. Attention mechanism is widely used in text classification tasks, and models based on self-attention mechanism like Transformer [21], BERT [22] achieves many exciting results on natural language processing tasks, such as machine translation, language modeling [23], and text classification. In view of the advantages of the attention mechanism mentioned above, we introduce attention mechanism into TextCNN to obtain the better performance.

2.4 Transformer and BERT

Transformer [21], BERT [22] achieves many exciting results on natural language processing tasks, such as machine translation, language modeling [23], and text classification. Transformer is a neural machine translation method with an attention mechanism, it is based solely on self-attention, without any recurrence or convolution operators. Bidirectional transformers (BERT) approach, which is one of the most powerful models in the NLP field. More lately, language learning networks such as Transformer-XL [23] and BioBERT [24] have further extended the Transformer framework.

However, the Transformer augments the input features by adding a positional embedding since the self-attention could not capture the positional information by itself. Despite its effectiveness on machine translation and language modeling, Transformer usually fails on the task with moderate size datasets due to its shortage of inductive bias.

3   Methods

3.1   Baseline Model: TextCNN

The textCNN model mainly uses a one-dimensional convolutional layer and a sequential maximum pooling layer. A one-dimensional convolutional layer is equivalent to a two-dimensional convolutional layer with a height of 1. The one-dimensional cross-correlation operation of multiple input channels is also similar to the two-dimensional cross-correlation operation of multiple input channels: on each channel, the core and the corresponding input is subjected to a one-dimensional cross-correlation operation, and the results between the channels are added to obtain the output result. The text matrix is composed of word vectors. The filter core sizes are 2, 3, and 4, respectively. After convolution pooling, the feature vector is obtained, and its dimension equals the number of convolution kernel sizes multiply by the size of each convolution kernel Number.

The calculation of TextCNN is as follows. Firstly, concatenating n words mapped into word vectors into one sentence. Then the convolution operation is carried out with multiple convolution kernels of different sizes to form feature map $\mathbf{c}$. The features of these convolutional filters are obtained by using 1-max pooling operation on $\mathbf{c}$. Finally, these features are concatenated and feed forward the softmax layer to get the tag probability distribution of the sentence.

$$c_i = f(\mathbf{w} \cdot \mathbf{x}_{i:i+h-1} + b), \tag{1}$$

$$\mathbf{c} = [c_1, c_2, ..., c_{n-h+1}] \tag{2}$$

$$y = softmax(1 - maxpooling(\mathbf{c})) \tag{3}$$

3.2   Optimized Components: GRU

Gated Recurrent Unit (GRU) is a variant of LSTM that has two gates. It combines the forget gate and the input gate into a single "update gate". It has also a "reset gate" and mixes the cell state and hidden state and some other changes. GRU has no additional memory cell to keep information, therefore, it can only control information inside the unit. The following formula represents the transformation process of information.

$$z_t = \text{sigmoid}(W_1 X_t + W_1 h_{t-1}) \tag{4}$$

$$r_t = \text{sigmoid}(W_2 X_t + W_2 h_{t-1}) \tag{5}$$

$$\tilde{h}_t = \tanh(W(r_t \otimes h_{t-1}) + W X_t) \tag{6}$$

$$h_t = (1 - z_t) \otimes h_{t-1} + z_t \otimes \tilde{h}_t \tag{7}$$

Here z_t is update gate, which decides for how much content or information is updated. r_t is reset gate, if the gate is set to zero, it reads input sequences and forget the previously calculated state. Further, $\tilde{h}_t$ shows the same functionality as in recurrent unit and h_t of GRU at time t represents the linear interpolation among the current $\tilde{h}_t$ and previous h_{t-1} activation states.

3.3 Optimized Components: Attention

Applying attention to the outputs of the convolutional layers served two purposes, it gave us insight about what filter sizes where more effective for the text classification task, which enabled to fine tune that parameter. For example, we tried large filter sizes and ended up eliminating that branch because it had a low attention score, and therefore it was somewhat a waste of parameters. More significantly, attention gives the network the freedom of assigning a score for each of the branches which boosted the model performance. The attention mechanism resulted in a higher accuracy compared to simply flattening or averaging the output of the convolutional layers.

The calculation of attention mechanism is as follows. Given a sequence of vectors $\mathbf{H} \in \mathbb{R}^{N \times D}$ where N is the length of the sequence and the D is the dimension of the vector. When doing attention, the module projects the H into three matrices: the query $\mathbf{Q}$, the key $\mathbf{H}$ and the value $\mathbf{V}$.

$$\mathbf{Q} = \mathbf{H}\mathbf{W}^Q, \mathbf{K} = \mathbf{H}\mathbf{W}^K, \mathbf{V} = \mathbf{H}\mathbf{W}^V \tag{8}$$

$$attention_i = \text{softmax}\left(\frac{\mathbf{Q}_i \mathbf{K}_i^T}{\sqrt{D}}\right) \mathbf{V}_i \tag{9}$$

where $\mathbf{W}^K, \mathbf{W}^K, \mathbf{W}^K$ are learnable parameters.

3.4 Model Architecture

In order to verify the impact of the convolutional layer vision on the three tasks of this project, We modified the convolutional layer, as shown in the following table. In the subsequent structural changes, We used the TextCNN-BS structure as the basis. In order to increase the impact of contextual information, We added the attention network to the TextCNN network. We use a Bi-GRU network or a attention network in the project to combine the impact of context on classification (Fig. 1).

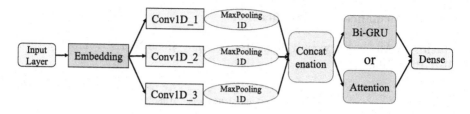

Fig. 1. Overview of model architecture

Table 1. Comparison of optimized method vs. baseline. TextCNN is set as baseline, and the filter core sizes are 2, 3, and 4, respectively. TextCNN-BS is one of the variants of TextCNN, which has different filter core sizes from baseline. TextCNN-SS's each filter core sizes is sets as 3. We use the TextCNN-BS to combine with Bi-GRU and Attention.

Network	Convolution layer size	After concatenate
TextCNN (Baseline)	2, 3, 4	Softmax
TextCNN-BS	3, 4, 5	Softmax
TextCNN-SS	3, 3, 3	Softmax
TextCNN-BS + BiGRU	3, 4, 5	BiGRU - softmax
TextCNN-BS + Attention	3, 4, 5	Attention - sigmoid

4 Experiments

4.1 Datasets

CAIL2018 contains more than 2.6 million criminal cases published by the Supreme People's Court of China, which are several times larger than other datasets in existing works on judgment prediction. Moreover, the annotations of judgment results are more detailed and rich. It consists of applicable law articles, charges, and prison terms. The length of the sentence includes 0–25 years, lifeless, and the death penalty. The data is stored in JSON format, each line is a piece of data, and each piece of data is a dictionary. In the use of the data set, We will use 64% as the training set, 16% as the validation set, and 20% as the test set.

The fields and meanings included in the data are: Fact: Fact description; meta: Annotated information, including: criminals: defendant (the data only contains one defendant); punish_of_money: fine (unit: yuan); accusation: count of crime; relevant_articles: relevant laws and regulations; term_of_imprisonment: sentence, sentence format (unit: month); death_penalty: whether it is a death sentence; life_imprisonment: whether it is indefinite; imprisonment: a term of imprisonment.

4.2   Performance on Classification

Platform: All models were training/testing on a single NVIDIA A100-PCIE-40 GB GPU (Table 2).

Table 2. Test accuracy on datasets

Network	Crime	Law	Prison term	Average
TextCNN (Baseline)	94.7	94.3	72.55	87.18
TextCNN-BS	95.1	95.3	70.7	87.03
TextCNN-SS	94.3	**95.6**	69	86.30
TextCNN-BS + BiGRU	95.3	94.8	73.65	87.92
TextCNN-BS + Attention	**95.7**	95.0	**76.75**	**89.15**

From the experimental results, we can get the following conclusions:

(1) In terms of overall performance, the TextCNN-BS+Attention and TextCNN-BS+BiGRU used in this project have achieved better results than the baseline, but the network with only the changed filter size has a slight decline in overall performance. The overall performance is the worst in the network using the same filter size, which proves that changing the receptive field is effective.
(2) In the single item analysis, the TextCNN-SS network using the same filter size has achieved the best results on articles of law recommendation, proving that the field of view affects the overall performance, but it is not the most effective for all situations.
(3) In Attention and bidirectional GRU, the joining of the Attention network exceeds GRU in all aspects.

4.3   Contrast of Parameters and Time

From the experimental results, we can get the following conclusions:

(1) In terms of overall performance, the parameters of TextCNN-BS+Attention model are significantly lower than other models, while ensuring less reasoning time consumption. It proves that the introduction of attention mechanism is effective and it maintains the real-time reasoning task while reducing the number of model parameters.
(2) In the single item analysis, for prison term prediction, the TextCNN-SS has the least model parameters, and the reasoning time is the same as that of TextCNN-BS+Attention model. However, it does not prove that the model has advantages, because according to the results of Table 1, the accuracy of this model in the prison term prediction is the lowest. In addition, TextCNN-BS+BiGRU has good performance, but lags behind all other models on inferring time (Table 3).

Table 3. Consumption of parameters and time

Network	Parameters (M)			Time for Infer (us)			FLOPs (M)
	Crime	Law	Prison	Crime	Law	Prison	Average
TextCNN (Baseline)	9.21	9.59	5.88	148	146	141	128.45
TextCNN-BS	9.40	9.79	6.08	149	144	137	131.20
TextCNN-SS	9.20	9.59	**5.86**	**136**	147	134	128.45
TextCNN-BS + BiGRU	7.57	7.59	7.48	955	947	985	106.43
TextCNN-BS + Attention	**6.52**	**6.51**	6.49	157	**139**	**132**	**47.42**

*The comparison of flops can also show the advantages of our method. Recently, the Transformer and its variants [22–24] has achieved excellent performance in various NLP tasks, but their flops have almost reached the range of 20000M to 40000M, such as Bert [22] reaching 21785M, which is completely different from our method.

(3) In conclusion, our proposed method including TextCNN-BS+BiGRU and TextCNN-BS+Attention, the joining of the Attention network exceeds BiGRU in all aspects.

5   Conclusion

In this project, we chose TextCNN as the baseline, verified the role of the receptive field in the classification of legal texts. We introduce a bidirectional GRU and an attention layer to enhance the impact of contextual information on classification. Experiments demonstrate that the combination of TextCNN and Attention achieves the better results than Bi-GRU for the comprehensive performance.

References

1. Hinton, G.E., Salakhutdinov, R.R.: Reducing the dimensionality of data with neural networks. Science **313**(5786), 504–507 (2006)
2. Nitin, J., Bing, L.: Review spam detection. In: 16th International Conference on World Wide Web, pp. 1189–1190. ACM (2007)
3. Ngai, E.W.T., Hu, Y., Wong, Y.H., et al.: The application of data mining techniques in financial fraud detection: a classification framework and an academic review of literature. Decis. Support Syst. **50**(3), 559–569 (2011)
4. Chu, Z., Gianvecchio, S., Wang, H., et al.: Detecting automation of Twitter accounts: are you a human, bot, or cyborg? IEEE Trans. Dependable Secure Comput. **9**(6), 811–824 (2011)
5. Caragea, C., McNeese, N.J., Jaiswal, A.R., et al.: Classifying text messages for the Haiti earthquake. In: 8th International Conference on Information Systems for Crisis Response and Management (ISCRAM) (2011)
6. Roitblat, H.L., Kershaw, A., Oot, P.: Document categorization in legal electronic discovery: computer classification vs. manual review. J. Am. Soc. Inf. Sci. Technol. **61**(1), 70–80 (2010)
7. Rajpurkar, P., Zhang, J., Lopyrev, K., et al.: SQuAD: 100,000+ questions for machine comprehension of text. arXiv preprint arXiv:1606.05250 (2016)

8. Xiao, C., Zhong, H., Guo, Z., et al.: CAIL2018: a large-scale legal dataset for judgment prediction. arXiv preprint arXiv:1807.02478 (2018)
9. Kim, Y.: Convolutional neural networks for sentence classification. In: Proceedings of the 2014 Conference on Empirical Methods in Natural Language Processing, pp. 1746–1751 (2014)
10. Battaglia, P.W., Hamrick, J.B., Bapst, V., et al.: Relational inductive biases, deep learning, and graph networks. arXiv preprint arXiv:1806.01261 (2018)
11. Zhou, J., Cui, G., Hu, S., et al.: Graph neural networks: a review of methods and applications. AI Open **1**, 57–81 (2020)
12. Kipf, T.N., Welling, M.: Semi-supervised classification with graph convolutional networks. In: ICLR (2017)
13. Huang, L., Ma, D., Li, S., Zhang, X., Wang, H.: Text level graph neural network for text classification. In: Proceedings of the Empirical Methods in Natural Language Processing, pp. 3442–3448 (2019)
14. Zhang, Y., Liu, Q., Song, L.: Sentence-state LSTM for text representation. In: Proceedings of the Meeting of the Association for Computational Linguistics, pp. 317–327 (2018)
15. Mihalcea, R., Tarau, P.: TextRank: bringing order into text. In: Proceedings of the 2004 Conference on Empirical Methods in Natural Language Processing, pp. 404–411 (2004)
16. Yao, L., Mao, C., Luo, Y.: Graph convolutional networks for text classification. In: Proceedings of the AAAI Conference on Artificial Intelligence, pp. 7370–7377 (2019)
17. Ding, K., Wang, J., Li, J., Li, D., Liu, H.: Be more with less: hypergraph attention networks for inductive text classification. In: Proceedings of the 2020 Conference on Empirical Methods in Natural Language Processing (EMNLP) (2020)
18. Liu, X., You, X., Zhang, X., Wu, J., Lv, P.: Tensor graph convolutional networks for text classification. In: Proceedings of the AAAI Conference on Artificial Intelligence, pp. 8409–8416 (2020)
19. Zhang, Y., Yu, X., Cui, Z., Wu, S., Wen, Z., et al.: Every document owns its structure: inductive text classification via graph neural networks. In: Proceedings of the Meeting of the Association for Computational Linguistics, pp. 334–339 (2020)
20. Bahdanau, D., Cho, K., Bengio, Y.: Neural machine translation by jointly learning to align and translate. In: Proceedings of ICLR (2015)
21. Vaswani, A., Shazeer, N., Parmar, N., Uszkoreit, J., Jones, L., et al.: Attention is all you need. In: NIPS, pp. 6000–6010 (2017)
22. Devlin, J., Chang, M., Lee, K., Toutanova, K.: BERT: pre-training of deep bidirectional transformers for language understanding. CoRR abs/1810.04805 (2018)
23. Dai, Z., Yang, Z., Yang, Y., Carbonell, J.G., Le, Q.V., et al.: Transformer-XL: attentive language models beyond a fixed-length context. CoRR abs/1901.02860 (2019)
24. Lee, J., Yoon, W., Kim, S., et al.: BioBERT: a pre-trained biomedical language representation model for biomedical text mining. Bioinformatics **36**(4), 1234–1240 (2020)

An Encryption Key Sharing Method for Contact Tracing Application

Yanji Piao[1], Zhijun Quan[2], and Xinghui Piao[3]([✉])

[1] Yanbian University, Yanji, Jilin, China
[2] China Agricultural University, Beijing, China
[3] Shanghai University, Shanghai, China
863263947@qq.com

Abstract. The world has seen many pandemics in the past. COVID-19, SARS, and H1N1 are some of them. During the period of epidemic prevention and control, tracing the source becomes a challenge to control the disease, and contact tracing applications are developed by many countries to slow down the spread of pandemics. However, the privacy problem is becoming one of the important issues in contact tracing systems nowadays. To protect the private information for infected persons and their potential contacts in the scenario which described in our paper, the effective encryption key sharing method can be applied to contact tracing systems. In this paper, we propose a key sharing mechanism for contact tracing application, it is allows a confirmed patient to hide their sensitive information from others when send the notification messages. Our mechanism is used to achieve such a user's privacy functionality. We present the security analysis and prove the security of the mechanism.

Keywords: Contact tracing · Privacy · Private information · Key sharing method

1 Introduction

In 2019, a newly emerged virus received widespread attention all over the world [7]. Later, WHO named it "corona virus desease-2019 (COVID-19)". As of May, 2022, the cumulative number of confirmed cases in China is more than 2300000, and worldwide is more than 528000000 [5]. One of the measures to deal with the spread of the COVID-19 is to identify close contacts of infected patients. Countries around the world have developed different contact tracing apps. The Australian government came up with CovidSafe [2], a contact tracing app based on Bluetooth technology. The person who is diagnosed as COVID-19 infected shares the stored information with the state health officials. We concluded 28 protocols and applications by reviewing previous literature which mentioned in our previous paper [11].

Although contact tracing applications have made great contributions to the control of the epidemic, the privacy problems are still going to be solved. The registration of contact tracking application must be accompanied by real name authentication. The leakage of information will lead to serious consequences, and the discrimination incurred by patients will also become a social problem.

© The Author(s), under exclusive license to Springer Nature Singapore Pte Ltd. 2023
S. Yang and S. Islam (Eds.): APWeb-WAIM 2022 Workshops, CCIS 1784, pp. 269–275, 2023.
https://doi.org/10.1007/978-981-99-1354-1_23

1.1 Applications

Contact tracing application was regarded as intra-group communication. Group communication is a relationship between three or more individuals who want to accomplish a common goal. The contact tracing of the diagnosed person or close contacts can also be said to be communication between communities. A community is a large group of interconnected life formed by several social groups or social organizations gathered in a certain field. Once someone is diagnosed, she/he needs to send a message to other communities and inform them that she/he has been diagnosed, so as to prevent the virus from spreading. The community leader is the group controller or group leader, and the other people are called group members. When a diagnosed person or close contacts send a notification message to the other community, the message must be encrypted with the secret key to prevent data leakage and the identity of the sender must be kept secret to hide identity. Moreover, only the sender and the members in target community will be able to recover the notification message.

1.2 Related Work

Singapore government technical agencies and health departments developed the first Bluetooth based contact tracking system—TraceTogether. In order to understand the working principle, Jason Bay et al. [8] released the privacy protection protocol Blue-Trace, which is used for developing TraceTogether. BlueTrace is a protocol for logging Bluetooth encounters between participating devices to facilitate contact tracing, while protecting the users' personal data [8]. The team from France and Germany [10] believes that the challenge of developing a contact tracking system is to build a solution that can not only protect privacy, but also resist malicious users and non malicious but curious authorities. Therefore, they proposed a privacy protection proximity tracking protocol ROBERT, which relies on the EU server infrastructure and temporary anonymous identifiers with security and privacy guarantees. PEPP-PT NTK (Pan-European Privacy-Preserving Proximity Tracing) protocol [9] is jointly created by a team of eight European countries. Its design fully conforms to the general data protection regulations issued by the EU, and can be used in EU countries through the anonymous joint mechanism. In the process of use, pepp-pt will not collect any personal identity information or location data.

DP3T (Decentralized Privacy-Preserving Proximity Tracing) agreement [4] is completed by teams from eleven European universities and scientific research institutions. DP3T protocol provides four privacy security measures: ensuring data minimization, preventing data abuse, preventing tracking users, and convenient application uninstallation. Apple [1] and Google [6] have developed the contact tracking system common to IOS and Android, and announced the detailed technical specification GAEN (Google Apple Exposure Notification) supporting its new privacy protection agreement. They are designed to provide users with a privacy friendly exposure notification.

The privacy protection protocols of the above five contact tracking systems all take Bluetooth as the core technology, and can be divided into centralized system architecture and decentralized system architecture according to different registration and information storage methods. The proportion of citizens' voluntary installation is still low, and it is impossible to play the efficiency of the contact tracking system to effectively prevent and control COVID-19.

China's personal health information code adopts GPS as the core technology, and combines GPS signals, base station data provided by telecom operators with QR code to track contacts. The difference from the European and American protocols is that China has adopted a highly centralized system architecture, and the epidemic prevention department has all the information of users. People have caused the public anxiety and doubt about the possible data privacy problems of personal health information codes. In response to such problems, the central cyberspace office, the general office of the National Health Commission and other departments have issued laws and regulations on personal information protection to strengthen supervision and prevent data leakage and abuse.

Once people are diagnosed with COVID-19 or their close contacts, private information may be leaked. Or the administrator of user data maliciously spreads private information. Therefore, in the post epidemic era, we need to design a system to collectively use big data to benefit society and protect every citizen by optimizing existing agreements. Based on this situation, we propose encryption key sharing method for contact tracing application.

2   Problem Statement

In this section, we describe the notations and the requirements to the proposed method.

2.1   Notations

In this section we shall list parameters we use throughout the rest of the paper.

1)   There are two communities named G1 and G2. G1 is the name of community 1, G2 is the name of community 2.
2)   GM1 is the leader of G1. GM2 is the leader of G2.
3)   V is a member ID $\in \{0, 1\}^{144}$ in G1.
4)   GK1 is the intra-group key of G1 shared between the members in G1 with GM1 securely in advance, pubGM1 is the public key of GM1, privGM1 is the private key of GM1.
5) $r \in \{0, 1\}^4$ and $R \in \{0, 1\}^4$ are random values generated by V.
6)   K is the encryption key between V and GM2.
7)   h() is the hash function, *key* is the encryption key between V and all the members in G2.

2.2   Requirements

We propose the following security requirements.

1)   To protect individual privacy, sender's identity is kept secret from others when a sender (a diagnosed person in community A) sends the notice message to the target community B in our scenario.

2) The sender gets signature from the signer without exposing his/her identity, in order to hide the identity.
3) With the aims of ensuring confidentiality, the system allows only the sender and the target community members to decrypt the secret information.

3   The Proposed Protocol

In this section, we will describe the method of sharing keys to satisfy the above requirements which is an extension of the ideas from our previous paper [12].

As shown in the Fig. 1 and Fig. 2, when the sender get sign from GM1, the identity of the sender is kept secret from GM1, in other word, GM1 cannot recognize who the sender is. However, GM1 can confirm the sender is one of his/her community member. Then, sender and GM2 share secret key securely. Figure 1 shows the sender how to hide the identity and get sign from GM1. Figure 2 shows that the sender never reveals who he/she is and gets an encryption key from GM2.

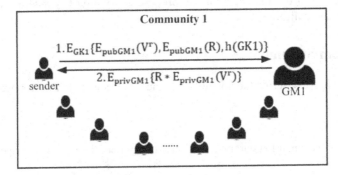

Fig. 1. Sharing between the sender and the community leader.

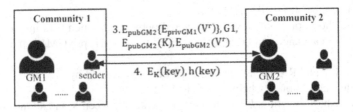

Fig. 2. Sharing between the sender and the target group.

1) The sender generates r and R randomly, r is used to hide the identity of the sender and R is used for increasing the information complexity. The sender encrypts $E_{pubGM1}(V^r)$, $E_{pubGM1}(R)$, $h(GK1)$ with GK1 and sends $E_{GK1}\{E_{pubGM1}(V^r), E_{pubGM1}(R), h(GK1)\}$ to his/her own community leader GM1. The sender's identity V is hidden by the random value r because of r power of V.

GM1 decrypts the message from the sender using GK1 and confirms from h(GK1) whether GK1 is tampered with others. Hence, GM1 determine whether the sender is one of his/her community member.

2) GM1 decrypts $E_{pubGM1}(V^r)$ and $E_{pubGM1}(R)$ using the private key privGM1 and obtain data V^r and R. The value V^r is saved in the database to judge whether there is the duplicated value in the same community. If GM1 judges that there is the same value V^r in G1, he/she notifies the sender to regenerate the random number r and repeats from the beginning step. However, we deem the space of V^r to be sufficiently large for collisions almost not to occur, even if there are a great many users. If the value V^r is unique in G1, GM1 encrypts the data V^r and R with the private key privGM1 and returns $E_{privGM1}\{R * E_{privGM1}(V^r)\}$ to the sender.

3) The sender decrypts $E_{privGM1}\{R * E_{privGM1}(V^r)\}$ with pubGM1 and gets $R * E_{privGM1}(V^r)$. Only the members who knows R^{-1} can derive $E_{privGM1}(V^r)$ from $R * E_{privGM1}(V^r)$. Since R is a random number generated by the sender, only the sender knows R^{-1}. The sender multiplies R^{-1} to $R * E_{privGM1}(V^r)$ and derives $E_{privGM1}(V^r)$. It makes sure the blind data V^r is signed by GM1.

4) The sender sends $E_{pubGM2}\{E_{privGM1}(V^r)\}$, G1, $E_{pubGM2}(K)$, $E_{pubGM2}(V^r)$ to the target community leader GM2, K is used for encrypting and decrypting the personal key share. GM2 decrypts $E_{pubGM2}\{E_{privGM1}(V^r)\}$ with his/her own private key privGM2 and derives $E_{privGM1}(V^r)$. At the same time, GM2 confirms the affiliated community of the sender. Then, GM2 derives the blind data V^r and the encryption key K by decrypting the message $E_{pubGM2}(V^r)$ and $E_{pubGM2}(K)$ with the private key privGM2. Although GM2 knows which community the sender belongs to, the identity of the sender has not been disclosed.

5) As mentioned our previous paper [12], GM2 calculates a key which will be shared between the sender and the community members in G2, and sends $E_K(key)$, h(key) to the sender.

6) The sender derives the *key* by decrypting $E_K(key)$ with K, and confirms the validity of the key using hash function h(key).

7) Finally, the sender encrypts the messages with the *key* and sends the encrypted messages to the target community members. It can only be decrypted by the sender in G1 and the target community members in G2 including GM2, and even GM1 cannot decrypt it.

4 Security Analysis

Privacy Protection: In our scenario, the identity of the infected person will be hidden when he/she send notification message to the target community because the ID of the sender is covered up with the random value r. In the protocol senders hide ID by calculating the power of r, it would thus be hard for the receivers in other community including leaders to track individuals by the IDs.

Ensure Uniqueness: There cannot be duplicate values in the community because the leader where the sender belongs to checks the uniqueness of V^r as mentioned in the Sect. 3. The person who has the same value V^r with infected person can eavesdrop on the notification messages or bring trouble to the communication.

Information Complexity: When the infected person gets signatures from his/her community leader, other persons in the same community are not able to intercept signature $(V^r)^{privGM1}$ of the leader because the random value R is generated by the infected person and only he/she can calculate R^{-1}. Furthermore, the random value r is used for hiding the identity of the infected person.

Integrity Validation: h(*key*) ensures that the *key* cannot be tampered with, thereby ensuring integrity of the messages. Once the sender receives $E_K(key)$ and h(*key*) from GM2, he/she calculate h(*key*) himself/herself and compares it with h(*key*) sent by GM2. If he/she achieves the same results, the key has not been modified by attackers, otherwise, the key has been modified.

Ensure Key Security: Since the *key* is encrypted with K which shares between the sender and the target community leader, even the leader of the sender cannot know the *key*.

5   Conclusion and Future Work

In this paper we have proposed encryption key sharing method that allows for privacy preserving contact tracing. Our method provides privacy for a user who is diagnosed with corona virus. Moreover, in Sect. 3 we briefly show how our method is secured, thus we believe it is of practical use in future situations where contact tracing is needed.

There are some deficiencies in our method, including the analysis of the overhead and the efficiency. We will consider integrating some encryption algorithms in recent years to improve the security and efficiency of the protocol in the future.

Acknowledgements. This research was supported by Social Science Research Project of Jilin Provincial Department of Education of China (NO. 612022062).

References

1. Apple: Privacy preserving contact tracing. https://www.apple.com/covid19/contacttracing. Accessed 30 June 2022
2. COVIDSafe. http://www.springer.com/lncs. Accessed 22 June 2022
3. Cui, D., Piao, Y.: A study on the privacy threat analysis of PHI-code. In: Gao, Y., Liu, An., Tao, X., Chen, J. (eds.) APWeb-WAIM 2021. CCIS, vol. 1505, pp. 93–104. Springer, Singapore (2021). https://doi.org/10.1007/978-981-16-8143-1_9
4. DP-3T. https://github.com/DP-3T. Accessed 30 June 2022
5. Epidemic Map. https://ncov.dxy.cn/ncovh5/view/pneumonia_peopleapp?from=timeline&isa ppinstalled=0. Accessed 29 May 2022
6. Google: Exposure notification api. https://www.google.com/covid19/exposurenotifications/. Accessed 30 June 2022
7. Hsia, W.: Emerging new coronavirus infection in Wuhan, China: situation in early 2020. Case Study Case Rep. **10**(1), 8–9 (2020)

8. Jason, B., et al.: BlueTrace: a privacy-preserving protocol for community-driven contact tracing across borders. Technical report, GovTech-Singapore (2020)
9. Pan-European Privacy-Preserving Proximity Tracing. https://github.com/pepp-pt/pepp-pt-documentation
10. ROBERT: ROBust and privacy-presERving proximity Tracing Claude Castelluccia. https://github.com/ROBERT-proximity-tracing/document. Accessed 22 June 2022
11. Paio, Y., Cui, D.: Privacy analysis and comparison of pandemic contact tracing apps. KSII Trans. Internet Inf. Syst. **15**(11), 4145–4162 (2021)
12. Paio, Y.-J., Kim, M.-J.: A study on the protection of consumers' personal information in online shopping. Acad. Soc. Glob. Bus. Adm. **15**(5), 209–223 (2018)

A Non-contact Interaction Method for 3D Display Devices

Mengkui Wang, Wentong Wang[✉], Feng Xu, and Qiang Liu

Academy of Artificial Intelligence, Beijing Institute of Petrochemical Technology, Beijing, China
wentongwang2022@163.com

Abstract. We propose a new interaction method to deal with the defects of traditional hardware in 3D interaction. The proposed method accurately captures human motion using the depth camera, obtains the RGB and depth images of the human body, and calculates the corresponding joints' 2D and 3D point coordinates. It estimates the human head orientation posture combined with the PnP function in OpenCV to control the presentation change of the scene in the three-dimensional display device. The method proposed in this paper is non-contact, does not need specific 3D glasses, and solves the problem that a monocular camera cannot be positioned. The experimental results show the precise control of the method proposed in this paper.

Keywords: Non-contact Interaction · Depth Camera · Head pose estimation

1 Introduction

In the past decades, people have become increasingly interested in the research and development of human-computer interaction devices [1]. Using a mouse and keyboard has become the standard for computer interaction [2]. More recently, touch-based interfaces in mobile phones, tablets, and touch-screen kiosks have been widely applied. While such equipment is more intuitive than a mouse or keyboard, the touch-screen size affects the cost of the device. If the interactive screen is huge, it would need other types of sensors to detect human posture for human-machine interaction. These sensors can be based on laser ranging scanners [3], depth cameras [4, 5], or IMU sensors.

Nowadays, non-contact interaction has been further developed and applied. The mainstream non-contact interaction is divided into gesture interaction [6] and head action interaction. However, both gesture interaction and head-action interaction have shortcomings, affecting the user experience to a certain extent.

Head pose is an essential indicator for studying human behavior and attention. When communicating with each other, people convey ideas by changing head postures, such as approval, disapproval, confusion, etc. Head pose estimation provides critical information for many human-computer interaction-related tasks, such as the steering of characters can be controlled by the head pose in VR interaction [7]. Nowadays, head pose estimation has become a hot topic in the field of computer vision and pattern recognition [8–11].

© The Author(s), under exclusive license to Springer Nature Singapore Pte Ltd. 2023
S. Yang and S. Islam (Eds.): APWeb-WAIM 2022 Workshops, CCIS 1784, pp. 276–284, 2023.
https://doi.org/10.1007/978-981-99-1354-1_24

In computer vision, the estimation of head pose refers to the use of computers to process and predict input pictures or videos through algorithms to determine the pose information of the character's head in three-dimensional space, namely pitch, roll, and yaw. A reliable algorithm for head pose estimation has the following characteristics: strong robustness, the capability to effectively handle occlusion, high real-time performance, and low computer resource rate [12].

In this paper, we propose a non-contact interaction method for 3D display devices, which uses a depth camera to obtain RGB and depth images of the human body without specific 3D glasses or any head-mounted devices, and adopts head pose estimation technology. For head pose estimation, the depth camera is used to accurately capture human movements, which solves the shortcomings of traditional contact interaction.

2 Related Work

2.1 Head-Based Interaction

The devices now widely used for head-motion interaction include VR (Virtual Reality) glasses [13], head display devices, etc.

VR technology, also known as virtual reality technology, is a computer simulation system that can create and experience virtual worlds. It uses computers to generate a simulated environment. The system emulation immerses the user in the environment. For example, HoloLens, Magic Leap One, etc. Use VR glasses technology to change the perspective of the scene in the glasses through the user's head movement. As a result, VR glasses are expensive, the general products are relatively bulky, and the user experience is not good.

The interactive software and hardware platform launched by zSpace [14] is a virtual reality system based on sensor-based 3D glasses to control scenes. Users can grab virtual objects and rotate or move them in 3D space using cameras, sensors, 3D glasses, and a stylus. The head-movement-based interaction method of zSpace requires using specific 3D glasses with sensors, which makes it not universal and expensive.

In this paper, we implemented a head-based interaction for arbitrary 3D glasses without extra sensors.

3 Mathematical Overview

The method proposed in this paper uses the depth camera to obtain the RGB-D image data of the human body. It estimates the corresponding joints' two-dimensional and three-dimensional point coordinates according to the derived RGB-D image. We then convert the 3D joint points from the camera coordinate system to the world coordinate system. The world coordinate system's origin is the nose tip. We calculate the rotation angle of the face plane in the world coordinate system from the neutral face position. The flowchart of the Program is given in Fig. 1.

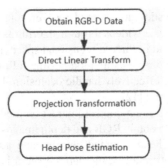

Fig. 1. The illustration of the proposed method's workflow

3.1 Obtain RGB-D Data

Usually, a controlled indoor environment needs to be established before acquiring image data: we place the depth camera above the computer screen, about 1.5 m from the ground; the user needs to be 1.0 m to 1.5 m in front of the depth camera; the user follows a pre-set usage protocol that includes slow head movements in the horizontal (yaw) and vertical (pitch) directions.

3.2 Direct Linear Transform

Through the RGB image and depth image, the two-dimensional and three-dimensional coordinates of the corresponding human joint points are estimated.

The coordinates of point p in the two-dimensional image are (x, y). The corresponding three-dimensional coordinates are (X, Y, Z). The translation formula from 2D image to camera 3D coordinate is shown in formula (1):

$$\begin{bmatrix} x \\ y \\ 1 \end{bmatrix} = s \begin{bmatrix} f_x & 0 & c_x \\ 0 & f_y & c_y \\ 0 & 0 & 1 \end{bmatrix} \begin{bmatrix} X \\ Y \\ Z \end{bmatrix} \tag{1}$$

where f_x and f_y are the focal lengths in the x and y directions, and (c_x, c_y) is the optical center, s is a scaling factor.

In this step, the coordinate transformation is performed from the camera coordinates (X, Y, Z) to the world coordinates (U, V, W). We use the equation in formula (2) in the camera coordinate system to calculate the position of point p.

$$\begin{bmatrix} X \\ Y \\ Z \end{bmatrix} = R \begin{bmatrix} U \\ V \\ W \end{bmatrix} + t \tag{2}$$

where $R \in \mathbb{R}^{3 \times 3}$ is the rotation vector and $t \in \mathbb{R}^{3 \times 1}$ is the translation vector.

3.3 Projection Transformation

Through translation transformation and rotation transformation, the three-dimensional coordinate system of the joint point is transformed from 2D image points to a 3D world coordinate system. The origin of the 3D world coordinate system is the nose tip point. The depth direction is transformed from the camera shooting direction to the nose tip direction.

As shown in Fig. 2, we select a plane α formed by the tip of the nose, the left eye, and the right eye, with the tip of the nose as the origin. When the posture of the human head changes, a new plane β will be generated. The acquired coordinates of the three points (the nose, left eye, and right eye) can determine the unique new plane. The angle between the two planes can be determined by calculating the angle between the two normal vectors of α and β. Vector $\vec{n}_1$ is the normal vector of β, and vector $\vec{n}_2$ is the normal vector of α. The formula for calculating the direct angle between two planes is shown in (3):

$$\theta = \arccos\left(\frac{|\vec{n}_1\vec{n}_2|}{|\vec{n}_1||\vec{n}_2|}\right), \theta \in [0°, 90°] \tag{3}$$

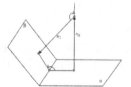

Fig. 2. α and β planes

The rotation axis is calculated according to the coordinates before and after the plane rotation, which is determined by the three points: the nose, the left eye, and the right eye. The plane where the rotation angle is located is the plane formed by α and β, and the rotation axis must be perpendicular to the plane.

The fixed vector of the nose before rotation, nose, left eye, and right eye is $a(a_1, a_2, a_3)$, and the corresponding vector after rotation is $b(b_1, b_2, b_3)$.

Suppose $\vec{a} = a_1i + a_2j + a_3k, +\vec{b} = b_1i + b_2j + b_3ka_3 \text{ k}, \vec{b} = b_1i + b_2j + b_3 \text{ k}$, formula (4) is obtained from the cross product definition:

$$\vec{a} \times \vec{b} = (a_2b_3 - a_3b_2)i + (a_3b_1 - a_1b_3)j + (a_1b_2 - a_2b_1)k \tag{4}$$

Thus, the rotation axis (n_x, n_y, n_z) is shown in formula (5):

$$\begin{pmatrix} n_x \\ n_y \\ n_z \end{pmatrix} = \begin{pmatrix} a_2b_3 - a_3b_2 \\ a_3b_1 - a_1b_3 \\ a_1b_2 - a_2b_1 \end{pmatrix} \tag{5}$$

According to the rotation axis and the rotation angle, the rotation matrix R_1 is calculated as shown in formula (6):

$$R_1 = \begin{bmatrix} r_{11} & r_{12} & r_{13} \\ r_{21} & r_{22} & r_{23} \\ r_{31} & r_{32} & r_{33} \end{bmatrix} \tag{6}$$

and the value of each item in the matrix is detailed in formula (7):

$$
\begin{aligned}
r_{11} &= \cos\theta + n_x^2(1 - \cos\theta) \\
r_{12} &= -n_z\sin\theta + n_x n_y(1 - \cos\theta) \\
r_{13} &= n_y \sin\theta + n_x n_z(1 - \cos\theta) \\
r_{21} &= n_z \sin\theta + n_x n_y(1 - \cos\theta) \\
r_{22} &= \cos\theta + n_y^2(1 - \cos\theta) \\
r_{23} &= -n_x \sin\theta + n_y n_z(1 - \cos\theta) \\
r_{31} &= -n_y \sin\theta + n_x n_z(1 - \cos\theta) \\
r_{32} &= n_x \sin\theta + n_y n_z(1 - \cos\theta) \\
r_{33} &= \cos\theta + n_z^2(1 - \cos\theta)
\end{aligned}
\tag{7}
$$

Rotation coordinate change: $C = R_1 * \vec{b}$, $C(c_1, c_2, c_3)$ is a new matrix obtained by multiplying the rotation matrix R_1 and the transformed coordinate b, and the newly obtained C is the rotated coordinate. The coordinate change calculation formula is shown in (8):

$$\begin{bmatrix} c_1 \\ c_2 \\ c_3 \end{bmatrix} = R1 \begin{bmatrix} b_1 \\ b_2 \\ b_3 \end{bmatrix} \tag{8}$$

3.4 Head Pose Estimation

We use the two-dimensional and three-dimensional coordinate points of six points (the nose point, left eye, right eye, left clavicle, right clavicle, and neck) to estimate the orientation of the human head.

The solvePnP function in OpenCV can solve the rotation matrix R_2 and the translation vector t_2, and use the two-dimensional and three-dimensional coordinates of six points (the nose tip, left eye, right eye, left clavicle, right clavicle, and neck) in the three-dimensional scene. The camera's internal parameter matrix can solve the absolute pose relationship between the camera coordinate system and the world coordinate system representing the 3D scene structure. The solvePnP algorithm solves the rotation matrix R_2 and translation vector t_2, as shown in (9).

$$(R_2, t_2) = \text{solvePnP}\left(\begin{pmatrix} x \\ y \\ 1 \end{pmatrix}, \begin{pmatrix} U \\ V \\ W \\ 1 \end{pmatrix}, \begin{bmatrix} f_x & 0 & c_x \\ 0 & f_y & c_y \\ 0 & 0 & 1 \end{bmatrix} \right) \tag{9}$$

The rotation matrix R_2 can be denoted as matrix formation in (10):

$$R_2 = \begin{bmatrix} r_{11} & r_{12} & r_{13} \\ r_{21} & r_{22} & r_{23} \\ r_{31} & r_{32} & r_{33} \end{bmatrix} \tag{10}$$

The head pose is represented by three Euler angles: Yaw, Pitch and Roll [15] as shown in Fig. 3. Thus, in order to calculate the head pose, three sets of the predicted landmarks are implemented to perform calculations of the three angles.

The corresponding Euler angles Pitch, Roll, and Yaw can be obtained from the rotation matrix R_2, as shown in formulas (11), (12), and (13):

$$Pitch = \text{atan2}(r_{32}, r_{33}) \tag{11}$$

$$Roll = \text{atan2}\left(-r_{31}, \sqrt{r_{32}^2 + r_{33}^2}\right) \tag{12}$$

$$Yaw = \text{atan2}(r_{21}, r_{11}) \tag{13}$$

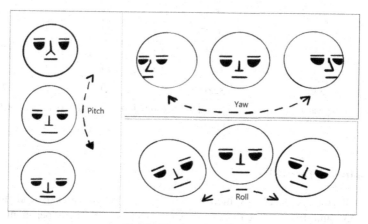

Fig. 3. Head pose angles.

4 Experimental Results Discussions

4.1 System Implementations

The prototype system uses the Kinect depth camera. The software is implemented using opencv3.4, C++. The running environment is Win10 OS system with Intel Xeon CPU.

Table 1. Experimental results

Head movement		Angle	Error	Pitch	Yaw	Roll
The units of the following values are°						
Horizontal	10°	10.3	0.3	12	20	20
	15°	15.47	0.47	12	9	18
	20°	21.02	1.02	15	22	26
	25°	24.67	0.33	11	6	30
	30°	30.6	0.6	12	13	32
	35°	34.86	0.14	10	22	32
	40°	40.6	0.6	16	16	32
	45°	46.22	1.22	16	25	47
	50°	50.55	0.55	6	25	49
	Average error	(0.3 + 0.47 + 1.02 + 0.33 + 0.6 + 0.14 + 0.6 + 1.22 + 0.55)/9 = 0.58°				
Vertical	5°	5.8	0.8	11	−3	11
	10°	10.25	0.25	18	−10	11
	15°	15.6	0.6	21	−2	11
	20°	20.73	0.73	22	−1	12
	25°	25.8	0.8	22	−11	12
	30°	30.46	0.46	22	−1	16
	33°	32.67	0.33	30	−2	2
	Aaverage error	(0.8 + 0.25 + 0.6 + 0.73 + 0.8 + 0.46 + 0.33)/7 = 0.56°				

4.2 Results and Discussion

The horizontal range of head movement is −50–50°, the vertical range of head movement is −33°−33°, and the effective distance range is 0.8 m−5 m. The experimental results are shown in Table 1.

The experiment is based on the posture of the human body, the angle formed by the human head and the vertical plane where the depth camera is located shall prevail during the test, in units of 5°/10°. More specifically, the head movement in the table refers to the head-turning angle of the human body. Angle represents the data of the vertical plane angle between the head and the camera calculated by the program. The error indicates the difference between Angle and Head movement. Pitch, Yaw and Roll refer to The Euler angle, and the average error refers to the average error of several data sets.

Before starting, we select a location with enough light and little interference. Then, we conduct the following steps in the experiment: 1. Fix the Kinect camera and adjust the optimal shooting angle. 2. After adjusting the head's angle, use a protractor to measure the angle between the head and the vertical plane. 3. When the data measured at the same angle varies greatly, the average value of three measurements is used to reduce the error.

Through the above experiments, we can see that the errors are relatively small. Therefore, we can use this method to control some applications through the gesture of the head pose.

5 Conclusion and Future Work

In this paper, we propose a non-contact interaction method for 3D display devices to address the defects of traditional hardware in 3D interaction. This method is non-contact and does not require specific 3D interaction glasses with unique sensors. Moreover, the proposed method has a simple structure, low time consumption, and can be applied to many face-related tasks.

As our method suffers from inaccurate capture in multi-person detection, we will further optimize it for better results and apply it to conduct multi-person interaction tasks.

Acknowledgement. This work was supported by the fund of the Beijing Municipal Education Commission, China, under grant number 22019821001.

References

1. Chunduru, V., Roy, M., Chittawadigi, R.G.: Hand tracking in 3d space using mediapipe and pnp method for intuitive control of virtual globe. In: 2021 IEEE 9th Region 10 Humanitarian Technology Conference (R10-HTC), Bangalore, India, pp. 1–6. IEEE Press, New York (2021). https://doi.org/10.1109/R10-HTC53172.2021.9641587
2. Karray, F., Alemzadeh, M., Saleh, J.A., Arab, M.N.: Human-computer interaction: overview on state of the art. Int. J. Smart Sens. Intell. Syst. 1(1), 137–159 (2017). https://doi.org/10.21307/ijssis-2017-283
3. Cheng, L., et al.: Registration of laser scanning point clouds: a review. Sensors 18(5), 1641 (2018). https://doi.org/10.3390/s18051641
4. Sharma, P., Joshi, R.P., Boby, R.A., Saha, S.K., Matsumaru, T.: Projectable interactive surface using microsoft kinect V2: recovering information from coarse data to detect touch. In: 2015 IEEE/SICE International Symposium on System Integration (SII), Nagoya, Japan, pp. 795–800, IEEE press, New York (2015). https://doi.org/10.1109/SII.2015.7405081
5. Shruti, M., Chittawadigi, R.G.: Offline simulation of motion planning of a planar manipulator in RoboAnalyzer and its integration with a physical prototype. In: 2021 IEEE 9th Region 10 Humanitarian Technology Conference (R10-HTC), Bangalore, India, pp. 1–6. IEEE Press, New York (2021). https://doi.org/10.1109/R10-HTC53172.2021.9641544
6. Alvissalim, M.S., Yasui, M., Watanabe, C., Ishikawa, M.: Immersive virtual 3D environment based on 499 fps hand gesture interface. In: 2014 International Conference on Advanced Computer Science and Information System, Jakarta, Indonesia, pp. 7–12, IEEE press, New York (2014). https://doi.org/10.1109/ICACSIS.2014.7065850
7. Cardoso, A., et al.: VRCEMIG: a virtual reality system for real time control of electric substations. In: 2013 IEEE Virtual Reality (VR), Lake Buena Vista, FL, USA, pp. 165–166, IEEE press, New York (2013). https://doi.org/10.1109/VR.2013.6549414

8. Hsu, H.W., Wu, T.Y., Wan, S., Wong, W.H., Lee, C.Y.: QuatNet: quaternion-based head pose estimation with multiregression loss. IEEE Trans. Multim. **21**(4), 1035–1046 (2018). https://doi.org/10.1109/TMM.2018.2866770

9. Patacchiola, M., Cangelosi, A.: Head pose estimation in the wild using convolutional neural networks and adaptive gradient methods. Pattern Recogn. **71**, 132–143 (2017). https://doi.org/10.1016/j.patcog.2017.06.009

10. Ruiz, N., Chong, E., Rehg, J.M.: Fine-grained head pose estimation without keypoints. In: 2018 IEEE/CVF Conference on Computer Vision and Pattern Recognition Workshops (CVPRW), Salt Lake City, UT, USA, pp. 2074–2083, IEEE press, New York (2018). https://doi.org/10.1109/CVPRW.2018.00281

11. Borghi, G., Fabbri, M., Vezzani, R., Calderara, S., Cucchiara, R.: Face-from-depth for head pose estimation on depth images. IEEE Trans. Pattern Anal. Mach. Intell. **42**(3), 596–609 (2018). https://doi.org/10.1109/TPAMI.2018.2885472

12. Luo, C., Zhang, J., Yu, J., Chen, C.W., Wang, S.: Real-time head pose estimation and face modeling from a depth image. IEEE Trans. Multimedia **21**(10), 2473–2481 (2019). https://doi.org/10.1109/TMM.2019.2903724

13. Ling, H., Rui, L.: VR glasses and leap motion trends in education. In: 2016 11th International Conference on Computer Science & Education (ICCSE), Nagoya, Japan, pp. 917–920, IEEE Press, New York (2016). https://doi.org/10.1109/ICCSE.2016.7581705

14. Sugimoto, M.: Augmented tangibility surgical navigation using spatial interactive 3-D Hologram zSpace with OsiriX and Bio-Texture 3-D organ modeling. In: 2015 International Conference on Computer Application Technologies, Matsue, Japan, pp. 189–194, IEEE press, New York (2015). https://doi.org/10.1109/CCATS.2015.53

15. Wlodarczy, M., Kacperski, D., Krotewicz, P., Grabowski, K.: Evaluation of head pose estimation methods for a non-cooperative biometric system. In: 2016 MIXDES - 23rd International Conference Mixed Design of Integrated Circuits and Systems, Lodz, Poland, pp. 394–398, IEEE press, New York (2016). https://doi.org/10.1109/MIXDES.2016.7529773

Author Index

S. Yang and S. Islam (Eds.): APWeb-WAIM 2022 Workshops, CCIS 1784, pp. 285–286, 2023.
https://doi.org/10.1007/978-981-99-1354-1

Printed in the United States
by Baker & Taylor Publisher Services